A Manual of Renal Transplantation

Richard DM Allen FRACS
Head, Transplantation Surgery Unit,
Westmead Hospital, Sydney

and

Jeremy R Chapman FRACP
Head of Nephrology, Westmead Hospital, Sydney

 SANDOZ

Presented with
the compliments of
Sandoz Pharma

Edward Arnold
A member of the Hodder Headline Group
LONDON   BOSTON   MELBOURNE   AUCKLAND

© 1994 Richard D. M. Allen and Jeremy R. Chapman

First published in Great Britain 1994

Distributed in the Americas by Little, Brown and Company,
34 Beacon Street, Boston, MA 02108

British Library Cataloguing in Publication Data

Allen, Richard
  Manual of Renal Transplantation
  I. Title II. Chapman, Jeremy R.
  617.4

ISBN 0–340–55154–2

Typeset in Palatino by Anneset, Weston-super-Mare, Avon.
Printed and bound in Great Britain for Edward Arnold, a
division of Hodder Headline PLC, Mill Road, Dunton Green,
Sevenoaks, Kent TN13 2YA by St Edmundsbury Press, Bury
St Edmunds, Suffolk and Hartnolls Ltd, Bodmin, Cornwall.

Preface

This manual has been written to provide a concise working guide to the practice of renal transplantation for resident medical staff, nurses, paramedical staff and students. We hope that these pages will provide an education to those in search of knowledge of renal transplantation; an introduction to those who may pursue a career in clinical transplantation; and a refuge for the resident and registrar faced with a clinical situation that they have not previously encountered. We have also attempted to provide a source of quick reference for those who need an answer or diagnostic pathway to an answer in renal transplantation.

We use the manual in our transplant unit and it has thus been modified by a combination of awareness of current literature, discussion, practical application and personal discovery. By the natural processes of medical advance, the 'immutable rule' of today is superseded by new didacticism. While we recognise this phenomenon is seen in the field of transplantation perhaps more quickly than in many other fields of medicine, we hope that this book can form the basis for your own manual of transplantation.

We would like to thank the many members of the Transplant Team and colleagues at Westmead Hospital who have provided the dedication and quality of care needed for the task of transplantation. We also wish to acknowledge the skills and tolerance of our own mentors, colleagues and patients who continue to teach us.

Richard Allen and Jeremy Chapman
Sydney, 1993

Acknowledgements

We would like to acknowledge the help and advice that we have received with this book from our colleagues: Gill Anderson, Peter Barclay, Tony Cunningham, Michael Earl, Kate Galton, Merle Greenberg, Simon Grunewald, David Harris, Elizabeth Hodson, Susan Lawrence, Yugan Mudaliar, Philip O'Connell, Peter O'Neill, Alan Sexton, Tania Sorrell and John Stewart.

We would like to acknowledge the free access that we have had to use data from the Australian and New Zealand Registry ANZDATA, the Collaborative Transplant Study and the University of California Los Angeles Transplant Registry, provided by Dr Alex Disney, Prof Gerhard Opelz and Dr Paul Terasaki respectively. The interpretations that we have placed on those data should not be regarded as representing the official policies and interpretations of the respective registries.

Invaluable secretarial skill provided by Mrs Robyn Hunt and Ms Gail McMullen, together with the artistic and graphical talents of Ms Kate Hilliger and Ms Tracey Pearl made this production possible. We also thank our wives for their support of this enterprise and our children for using the reams of discarded paper perhaps more productively than we did.

Contents

1 Organisation of renal transplantation

In the space of little more than 30 years, renal transplantation has progressed from an essentially unsuccessful experimental procedure conducted in the isolation of a small number of pioneering medical research centres to a widespread and routine procedure for an estimated 35,000 patients worldwide each year. Progress was achieved by necessity and one might speculate that few patients would have been prepared to embark upon the hazardous experimental path by which renal transplantation developed had haemodialysis always been a viable and safe long term treatment for renal failure. Indeed, the transplant recipients of today are indebted to the courage of renal transplant recipients of the 1950s and 1960s, as well as to the vision and perseverance of those who treated them.

Despite providing the best quality of life and the most cost effective treatment option for treatment of endstage renal disease, renal transplantation remains at the challenging and exciting interface between clinical research and laboratory science. Its development has also been the catalyst for transplantation of other solid organs, the practice of immunosuppression and much of clinical immunology. Nevertheless, failures, both acute and long term, continue to serve as a salutary reminder of our incomplete understanding and control of the human immune system. Furthermore, the very success of renal transplantation has brought a new challenge for the 1990s in the form of donor organ shortage and possible commercialisation of the donor supply.

A brief history

In the early part of this century, four surgeons, Ullman from Austria, Unger from Germany, together with Jaboulay and Carrel from France, demonstrated in large animal experimental models that transfer of organs from one individual to another and from animal to man was technically feasible. Each was of the opinion that factors other than thrombosis were responsible for the eventual graft loss. Between 1933 and 1949, a Ukranian surgeon, Yu Yu Voronoy, treated six patients with acute renal failure by transplantation of a human kidney into the recipient's thigh. No graft function was obtained, principally because Voronoy did not recognise the deleterious effect of six hours of warm ischaemia. In 1946, Hufnageal in Boston, USA achieved renal function for three days after transplantation of a cadaver kidney into the arm of a patient with acute renal failure. A series of kidney transplant procedures without the use of significant immunosuppression were subsequently reported from Boston but without any long term success.

The modern era of renal transplantation was heralded by the successful transplantation of a kidney between identical twin brothers in Boston in 1954. The transplanted kidney continued to function for eight years with eventual failure due to recurrent primary renal disease. Clearly this success was of limited practical value as very few renal failure patients had identical twins to provide donor kidneys. It did, however, prompt pioneering groups in Boston and Paris to perform unrelated transplants using total body irradiation. Unfortunately, this very non-specific mode of immunosuppression proved to be both cumbersome and dangerous, with an unacceptable mortality from overwhelming infection.

The immunosuppressive properties of the purine analogue 6-mercapto-purine were first noted by Roy Calne in dog renal transplants in 1960. Azathioprine, which was developed as a similar but less myelotoxic agent, also suppressed renal transplant rejection in dogs and subsequently became available in clinical practice. When azathioprine was used in combination with corticosteroids, renal transplantation finally became a viable treatment of endstage renal disease, with graft survival of about 50% at one year. Some of these unrelated grafts are still functioning nearly 30 years later. Results were further improved by routine use of lymphocytotoxic crossmatching of the donor and recipient prior to transplantation; prospective matching of histocompatability antigens; the use of antilymphocyte preparations; and deliberate blood transfusion policies (Table 1.1).

The introduction of cyclosporin into clinical practice in 1978 by Roy Calne increased the chances of graft survival by up to 20% compared with immunosuppressive therapy with azathioprine and corticosteroids. The 1980s saw the first clinical use of monoclonal antibody therapy and one year first cadaver graft survival of between 85 and 90%.

An integrated treatment programme

Renal transplantation is only one of several options for replacement renal therapy, with haemodialysis or peritoneal dialysis often required before, during or after transplantation. For this very practical reason, renal transplant units have usually developed as an integrated part of a wider renal failure treatment programme. It is nevertheless argued by many that the more successful transplant units function with a decision making hierarchy that is independent of that involved with dialysis treatment. Increasingly, the care of transplant recipients has become the domain of specialist transplant clinicians trained in the management of patients with immunosuppression related problems of infection, cardiovascular disease,

Table 1.1   Milestones in the history of renal transplantation

1902	First successful experimental (canine autograft) transplantation
1912	Alexis Carrel awarded Nobel Prize for describing techniques of organ grafting
1933	First human allograft kidney transplant performed by Yu Yu Voronoy, in USSR
1943	Rejection observed by Medawar and Gibson to be a result of active immunisation
1946	First functioning human allograft performed, in Boston, USA
1954	First long term surviving human renal isograft performed, in Boston, USA
1960	First long term surviving immunosuppressed allograft performed, in Paris, France
1960	Observation of effectiveness of 6-mercaptopurine in canine renal transplantation
1962	Widespread use of azathioprine and steroids in renal transplantation
1967	Machine hypothermic perfusion of kidneys providing 72 hours of storage
1967	Clinical introduction of antilymphocyte globulin for prophylaxis against rejection, in Denver, USA
1968	Brain death criteria determined by ad hoc committee of Harvard Medical School
1968	First prospective histocompatibility locus antigen matching of cadaver kidney recipients, in Melbourne, Australia
1969	Cold storage of kidneys for 24 hours
1978	First clinical use of cyclosporin, in Cambridge, England
1981	First clinical use of mouse monoclonal anti-CD3 antibody (OKT3)
1990	First clinical use of FK506, in Pittsburgh, USA

malignancy and perhaps, declining renal function! Others, however, argue with some justification that the increasing specialisation of transplantation impairs the continuity of patient management that can be provided by an appropriately trained nephrologist.

The organisational relationship of dialysis and transplantation is by necessity, often determined by the size of the treatment programme. Irrespective of size, the complexity of the challenge of treating endstage renal disease requires cooperation between surgeons, physicians and laboratory scientists. The economic reality of this cooperation in the 1990s dictates that firstly, both where and when feasible, all patients should be considered for renal transplantation; and secondly, donor kidneys, a limited resource, should be transplanted in the best of possible circumstances.

Dialysis

In a world of perfect renal transplantation, dialysis units would be obsolete. All patients with endstage renal failure would be suitable for transplantation and would receive a kidney before the need for dialysis. Furthermore, all transplanted kidneys would have primary renal function and would not fail for either technical or immunological reasons. In the real world, patients often present late and require dialysis, at least in the short term, to prepare them better for the transplant surgical procedure. About 20% of transplanted kidneys do not work immediately; about 15% have failed by the end of the first year of transplantation, and a further 3 to 5% are lost each year thereafter from either chronic rejection or death of the recipient with a functioning graft. Finally a period of dialysis is almost inevitable if cadaver donor kidneys are to be allocated on the basis of histocompatibility matching rather than priority based on the time on the transplant waiting list. Dialysis thus remains as important in the 1990s as it was in the 1960s.

Dialysis is a very expensive form of treatment, the cost of which may preclude its widespread use in some countries. The cost of one year of haemodialysis or peritoneal dialysis is similar to that of a renal transplant procedure in the first year. Thereafter, the cost of renal transplantation is limited to the cost of immunosuppressive agents and is about 20% of the cost of dialysis per year. More importantly, however, the quality and quantity of life for dialysis patients are both restricted when compared with transplantation.

Size of the transplant unit

The median sized renal transplant unit in the USA performs about 30 transplants per year, similar to that in Australia but less than that of the majority of units in Europe. Although there are no firm guidelines for a minimum size of a transplant unit, larger units have both logistic and economic advantages, even though graft survival in many smaller units may be comparable to that of larger units. A separate ward area for nursing transplant recipients becomes feasible. Provision of skilled staff, particularly for 24 hour cover of the unit, also becomes practical as the size of the unit increases. The less tangible benefit of a large unit is the environment created by people working towards a common goal, especially when associated with an active laboratory research programme. The activity associated with larger units also enhances the level of experience of associated specialist groups within the hospital, such as those involved in imaging the renal transplant or in treating immunosuppression related infections.

The origin of donor kidneys

The success and safety of renal transplantation has provided a demand for donor kidneys which has not been met in most communities. Living donor kidney donation can be undertaken with acceptable risk, but is of limited application and avoided by some transplant units. Cadaver kidney donation relies upon the certainty of diagnosis of brainstem death, a diagnosis made by those with no involvement in transplantation.

Coordination of organ donation

Separation of the two clinical practices of the diagnosis of death and the use of donated organs for transplantation has led to development of the professional transplant coordinator and, particularly in the USA, to quite large organisations devoted to organ procurement. The transplant coordinator acts as an intermediary between the donor hospital and the transplant team, increasing the distinction between death and donation as separate events, and providing for development of expertise in the various organisational and legal activities surrounding organ donation. There is, however, a temptation to rely on the multiskilled transplant coordinator for expertise spanning such disparate activities as operating room assistant, perfusionist, public relations officer and bereavement counsellor. When financial support permits and where transplant coordinators serve large populations, organ procurement organisations are able to develop these separate components of transplant coordination more successfully.

Donor hospital support

Very few transplant units rely solely on the hospital in which they are based for organ donors. In hospitals with large trauma and neurosurgical services, 2–3% of deaths may be organ donors if consent is gained. Support by the transplant units of those hospitals is paramount and includes timely and efficient donor surgery as well as counselling staff involved in caring for donors. The delicate and difficult task of ensuring that donor hospitals needs are met is usually delegated to the transplant coordinator. In smaller donor hospitals the transplant unit may have a role in counselling and supporting the families of donors. Larger hospitals may have their own established mechanisms for helping bereaved families, irrespective of whether brain death and organ donation are involved. It is thus important for transplant units to be sensitive to and responsive to this spectrum of needs.

Public education

Public opinion surveys usually show that almost everyone has heard of transplantation (98%) and most are in favour of donation for themselves (60 to 70%). A minority are prepared to donate the organs of a family member (35%) or actually to sign a donor card or driver's licence (30 to 50%). Very few people understand brain death, and many believe that organ donors are determined at the roadside after a motor vehicle accident. Even amongst qualified medical and nursing staff working in intensive care or accident and emergency departments, the proportion that can explain brain death correctly is alarmingly low (30%).

There are two possible approaches to this widespread ignorance. The first is to legislate that organ donation will take place in the event of brain death. The second approach is to educate everyone, which has the disadvantages of cost and enormity of the task, but nevertheless may be more acceptable to a democratic community. In Australia, the task has been placed primarily in the school room of 14 to 16 year old children, with an 'education package' developed by the Australian Kidney Foundation. Until these children reach adulthood or succeed in the task of educating their parents, public information and marketing techniques are being used to varying effect.

Data collection

Transplantation, perhaps because it is a comparatively recent treatment modality, is one of the few medical activities where the total experience of geographic regions, individual countries or groups of countries is known.

Outcome measures are therefore available to those involved in transplantation, public health policy or hospital administration. Contribution of data to registries is a creed in transplant units without which the slow but steady improvement in outcomes would almost certainly not have occurred.

Transplant registries

In the 1960s, when relatively few renal transplants were being performed, it was important to collect data from many centres to analyse the results and determine the most significant influences on outcome. The University of California at Los Angeles (UCLA) established a registry, predominantly but not exclusively of units in the USA. Many influential reports have come, since then, from the UCLA Registry which now publishes an annual volume on clinical transplants edited by Paul Terasaki. The United Network for Organ Sharing (UNOS) is a more recent and complete registry of transplant activity in the USA. In 1982, Gerhard Opelz created an International Collaborative Transplant Study (CTS) based in Heidelberg, Germany. In its tenth year, 308 centres had contributed data on a total of 113,864 renal transplants. The CTS has thus become a powerful registry tool for analysis of transplant outcome variables, where even small subsets of the data carry sufficient statistical power to determine differences or lack of difference between alternatives. In Australia and New Zealand, the ANZDATA Registry has records of every patient ever treated for endstage renal disease by either dialysis or transplantation. The registry is based in the renal unit at Queen Elizabeth Hospital, Adelaide, and produces an annual report of all Australian and New Zealand activity edited by Alex Disney.

Data from each of these registries have been reproduced throughout this manual to support current clinical practice or to demonstrate clinically important effects. In doing so, we acknowledge both the organisers and staff of those registries together with the hundreds of members of participating renal units.

The 'centre effect'

One of the consequences of multiple transplant units or centres contributing to the various registries is that individual units can be compared with each other. In the UCLA Transplant Registry, transplant centre variation is the most influential factor in early graft survival. This poses the registries with their most delicate task, notably reporting of centre specific data in such a way that those who contribute do not withdraw. All registries have found that some transplant units have clearly better results than others, an example of which is apparent in Figure 1.1. Interpretation of data, however, is often difficult because of statistical variation and the variety of patients transplanted. To date, health economists, funding agencies, health insurers and patient groups have reacted only slowly to differences in transplant unit graft survival.

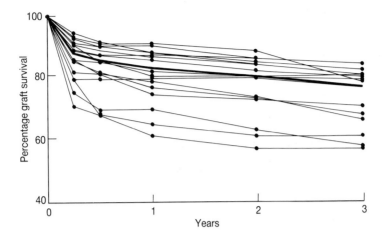

Figure 1.1 First cadaver graft survival for kidney transplant since 1987 in 13 Australian transplant units. The solid line represents the Australian mean graft survival. (Source: ANZDATA Registry)

Further reading:

1. Diethelm AG. Ethical decisions in the history of organ transplantation. *Ann Surg* 1990; **211**, 505–520.
2. Brent L, Sells RA. Notes on the history of tissue and organ transplantation. In: *Organ transplantation: Current Clinical and Immunological concepts. Brent L, Sells RA (eds).* London: Bailliere Tindall, *1989.*
4. PI Terasaki (ed). *History of Transplantation: Thirty Five Recollections.* Los Angeles: UCLA Tissue Typing Laboratory, 1991.
5. Starzl TE. *The Puzzle People. Memoirs of a Transplant Surgeon.* Pittsburgh: University of Pittsburgh Press, 1992.
6. Hume DM, Merrill JP, Miller BF, Thorn GW. Experiences with renal homotransplantation in the human. *J Clin Invest* 1955; **34**, 327–382.
7. Gjertson DW, Terasaki PI. The large center variation in half-lives of kidney transplants. *Transplantation* 1992; **53**, 357–362.
8. Evans RW, Manninen DL, Dong F. The center effect in kidney transplantation. *Transplantat Proc* 1991; **23**, 1315–1317.

2 The recipient

Careful assessment of potential renal transplant recipients both by those who are responsible for their long term care and by the team responsible for transplantation is an essential mechanism for all renal units. The patient needs both to consider the options available and to meet those offering transplantation. The process of assessment provides these opportunities and is critical to a successful outcome of transplantation.

General issues

Renal transplantation is not a trivial undertaking on the part of either patient or transplant centre. The initial perceptions that the patient has about renal transplantation will be influenced by television coverage of successes, meeting patients who may have had good or bad experiences, and by their optimism that it offers a route back to normality from chronic renal failure and dialysis treatment. The net result is that few patients have a balanced view of the advantages and disadvantages of transplantation and very few have an accurate understanding of the postoperative implications.

Education

Assessment of a potential recipient and their education about transplantation thus go hand in hand, so that the transplant centre can evaluate their suitability and the patient can decide if they want to accept transplantation. Education of the patient implies that they discuss transplantation with their doctors, transplant specialists, nurses and transplant coordinators. Reading matter and videos designed to teach patients about the realities allow the issues to be considered at home with the family. Part of this process will involve introducing the concept of living related donation to both the patient and family. Centres vary in their policies towards living donation, as discussed in Chapter 4, some requiring patients to present with a living related donor and some refusing to undertake living related donation. It is a responsibility of the centre to demonstrate the range of alternatives available.

Age

Age of the recipient has been taken as an indicator of risk. In the past, strict application of a cut-off age, above which transplantation will not be considered, has been used to direct scarce organ supply to younger patients. If the risk of death from transplantation is relatively high, patients have to evaluate that against the disadvantages of continued dialysis. Age is not in fact the criterion defining risk for anaesthesia, since it is on the whole the symptomatic and asymptomatic co-existing diseases which determine this risk. Age itself may be used as a general indicator of that risk, leading to patients over 55, 60, 65 or 70 years being refused. This argument is based upon what is best for the individual patient. The second set of arguments is based upon what is best for the community as a whole, or in other words how the maximum benefit can be derived from the scarce resource. If the patient's chance of survival is less than the graft's, this would lead to wasted years of graft function. It might thus be seen as logical not to transplant patients with less than a 50% chance of surviving for ten years. Age becomes, most naturally, the primary focus of this decision and leads some centres to place either barriers to transplantation over the age of 60 years, for example, or relative disadvantage in allocation systems so that a younger patient will receive a particular kidney if all other selected factors, such as HLA match and time waiting, are equal.

Data in Table 2.1 show the status of patients treated by dialysis in Australia with respect to absolute or relative barriers to transplantation. Of those treated by dialysis between 55 and 64 years, 28% were accepted for transplantation but only 5.5% of those over 65 years were on the waiting list.

Renal disease

The patient's primary renal disease has implications for transplantation in a number of instances either because of risk of recurrence of the disease in the graft, the possibility of infection arising as a result of the primary renal disease, or because technical difficulties with the operation may be predicted. Diabetes mellitus, with its numerous complex implications, is considered separately in Chapter 16.

Recurrent disease

The risk that the patient's primary renal disease will destroy the transplanted kidney varies from near certainty in the case of Goodpasture's syndrome with circulating antiglomerular basement membrane antibody, to very low in the absence of those antibodies. Disease recurrence in renal transplants, with loss of the graft, are factors that have to be considered in the diseases listed in Table 2.2.

In practice the rate of histological recurrence and loss of grafts from recurrent disease is not significant enough to inhibit transplantation except in three situations. Firstly in Goodpasture's syndrome, as mentioned above, anti-GBM antibodies must be absent for one year before transplantation. Plasmapheresis has the benefit of shortening the duration of anti-GBM

Table 2.1   Status of patients on dialysis in Australia with respect to the Transplant Waiting List

	HD	PD	Total	
Awaiting transplantation				
Fit non-sensitised	589	267	856	(31%)
moderately sensitised	69	21	90	(3%)
highly sensitised	142	28	170	(6%)
excellent match requested	35	2	37	(1%)
Not awaiting transplantation				
Patient refusal	97	54	151	(6%)
Technically difficult	38	2	40	(1%)
Recurrent nephritis risk	10	1	11	(<1%)
Multiple graft failure	17	2	19	(<1%)
Age				
Co-existing disease:				
respiratory	24	11	35	(1%)
cardiac	108	66	174	(6%)
malignancy	55	30	85	(3%)
cerebral	9	2	11	(<1%)
peripheral vascular	9	17	26	(1%)
hepatic	8	0	8	(<1%)
infection	18	2	20	(1%)
unspecified	109	46	155	(6%)
Obesity	13	3	16	(<1%)
Other	25	17	42	(2%)
Awaiting assessment	90	48	138	(5%)

HD = Haemodialysis
PD = Peritoneal dialysis
Data from the ANZDATA Registry 1990

antibodies and thus may be used solely for this indication in some patients.

Secondly, primary oxalosis is rare, but loss of the graft and patient is common after renal transplantation. Data reported to the European Dialysis and Transplant Association Registry on a total of 221 patients, 98 of whom received first grafts, showed patient survival of 74% at three years, but with three year graft survival as low as 17% for first cadaveric grafts. Recurrence of primary oxalosis accounted for about one third of the graft losses. The solution gaining increasing favour is to transplant both the liver and kidney, thereby solving both the metabolic disease and renal failure at the same time.

Finally, focal and segmental glomerulosclerosis (FSGS) has a high recurrence rate, especially if the initial histology had a significantly proliferative mesangium and if the time from initial diagnosis to endstage renal failure was short. In these circumstances the recurrence rate may be as high as 75% and the failure rate nearly as dramatic. One must certainly be reluctant to embark upon living related donation in this situation, though after careful forewarning, cadaveric transplantation is often undertaken. In systemic lupus erythematous, on the other hand, recurrent disease is uncommon after transplantation. This may either be because the disease has 'burnt out' by the time the patient comes to transplantation, or because of the coincidental effect of immunosuppression.

In the short term, recurrent disease has been a relatively rare cause of graft loss. It must, however, be anticipated that as loss from acute

Table 2.2 Recurrent primary renal disease in renal transplants

Disease	Literature survey*			ANZDATA		
	n	% recurrence	% failure	*n*	% recurrence	% failure
Glomerulonephritis						
Focal and segmental glomerular sclerosis	155	23	11	263	6.5	5
IgA	34	53	3	286	<1	<1
Henoch Schönlein purpura	39	31	8			
Mesangiocapillary Type I	140	30	9	176	4.5	3.5
Mesangiocapillary Type II	61	88	10	52	15	13.5
Haemolytic uraemic syndrome	–	–	–	36	14	14
Goodpasture's syndrome	30	5	<1	89	2	2
Crescentic–idiopathic GN	18	33	11	–	–	–
Membranous	22	55	0	120	0	0
Oxalosis				6	80	80

Notes

* Literature survey: Cameron JS. *Transplantation* 1982; **34**, 237–245
– Data not available

There are intrinsic differences in the way these data arise. In the literature survey, cases and clusters tend to be reported preferentially, while in the ANZDATA survey only definite cases of recurrence, proven histologically in both native kidneys and transplants, are included.

Table 2.3 Diseases requiring pre-operative investigation of ureter, bladder and urethra anatomy and function

Bladder	Prune belly syndrome
	Neurogenic bladder
Urethra	Urethral valves or stricture
	Bladder neck obstruction
	Prostatic enlargement
Ureter	Vesico-ureteric reflux
	Uretero-vesical stricture
	Ureteric obstruction
	Pelvi-ureteric junction stenosis

and chronic rejection declines, and as the natural histories of recurrent glomerulonephritis unfold, the patient's primary disease will become of greater importance.

Primary renal diseases with technical implications

Polycystic kidney disease   Adult polycystic kidney disease may lead to a number of problems after transplantation that are predictable from the history and examination. Recurrent cyst infection, haemorrhage into cysts or haematuria, and severe hypertension may all be gleaned from the history and provide indications for either unilateral or bilateral nephrectomy. Physical bulk of the kidneys may leave no space for a conventionally placed renal transplant or may mandate the side on which a transplant can be performed, unless unilateral or bilateral nephrectomy is undertaken first.

Urological disease   Careful pre-operative assessment of the bladder and urethra are needed in some but not all patients (Table 2.3). When the bladder is known to be abnormal a micturating cystogram is useful, together with urodynamic assessment of urine flow. Surgical correction of obstruction and either enlargement of the bladder with a caecoplasty or creation of a urinary conduit may be needed. In the case of severe bilateral reflux with continuing urinary tract infection, bilateral nephrectomy is now a better course of action than risking recurrent graft pyelonephritis, given the availability of erythropoietin to treat the resulting anaemia.

Co-existing disease

Only a few patients have an uncomplicated renal disease that does not involve any other systems. Most patients either have multisystem diseases involving the kidneys together with other organs, or have developed co-existing disease as a result of hypertension, hyperlipidaemia or uraemia. It is therefore standard procedure to assess not only the patients' renal disease but also their symptomatic and asymptomatic disease elsewhere. A list of routine pre-operative investigations is shown in Table 2.4. In addition to this list, further investigation may be indicated in particular situations as discussed below.

Table 2.4   Pre-operative assessment of all potential renal transplant recipients

1. *History*
    - Full history of renal disease and knowledge of investigations performed
    - Full history of previous medical history including previous blood transfusions
    - Medications
    - Allergies
    - Family and social history
    - Smoking, alcohol and other drug abuse

2. *Physical examination*
    - Full clinical examination, recording all abnormalities including central nervous system
    - Particular attention to:
      peripheral vascular system   } to determine technical
      abdominal examination   } suitability for transplantation
    - Dental examination

3. *Routine investigations*
    Blood   Full blood picture, blood group, red cell antibody screen
                   Urea, creatinine, electrolytes, calcium, PO_4, liver function tests, uric acid
                   Fasting glucose, triglycerides, cholesterol
                   Viral studies: HIV, hepatitis B, hepatitis C,
                                    CMV, herpes varicella/zoster,
                                    EBV, herpes simplex
    Chest X-ray
    Urine microscopy and culture
    Electrocardiogram

Technical barriers to transplantation surgery

Obesity Obesity may make surgery either impossible or unreasonably risky and acts as a significant relative contraindication to transplantation in patients more than 15 or 20% above their ideal weight. Patients treated by haemodialysis may, by dint of personal effort and discipline, lose sufficient weight. This is not as straightforward in a number treated by CAPD since the mandatory glucose load, especially in those losing peritoneal ultrafiltration capacity, can counteract any tendency to weight loss. Patients using CAPD should, if doubt exists, be examined without intraperitoneal fluid, to assess the abdomen with accuracy.

Peripheral vascular disease Peripheral vascular disease commonly creates minor and occasionally major problems with transplantation surgery. It is therefore important to assess the likelihood of severe atherosclerotic disease in the iliac vessels by examining the femoral and distal pulses and investigating with Doppler ultrasound and angiography when there is evidence of impaired blood flow.

Anaesthetic and early postoperative risks

Cardiac disease There are many approaches to detection of cardiac disease, of which ischaemic heart disease poses the greatest problems. Routine assessment of cardiac valve pathology including history, physical examination, chest X-ray and echocardiography will allow decisions on operative risk and requirement for prophylactic antibiotics. Patients with symptomatic ischaemic heart disease or abnormal ECG need careful investigation and management, with coronary angiography and artery bypass grafting as indicated.

The most difficult decision to make is the place of routine invasive investigation in patients with no symptoms, a normal ECG and normal CXR. The various alternatives are; routine coronary angiography for all patients; routine thallium-persantin, exercise thallium cardiac scanning or echocardiogram, proceeding to coronary angiogram in those with positive results; or no investigation. It is, however, possible on the basis of age, history of hypertension and other risk factors, and co-existing disease such as diabetes mellitus, to define a group of asymptomatic patients in whom those routine tests are more likely to yield dividends. Many centres would therefore adopt this middle course of extensively screening the higher risk patients only.

Respiratory disease The criteria for acceptance of patients for renal transplantation are no different to those of other operations of similar length and complexity. Two respiratory diseases need particular assessment, however; pulmonary tuberculosis and bronchiectasis. Tuberculosis needs to have been treated effectively before elective immunosuppression, which must then be accompanied, for at least six months, by prophylaxis against recrudescence of the tuberculosis. Bronchiectasis may, on the other hand, present a long term challenge in an immunosuppressed patient. If the bronchiectatic lesion is localised and recurrent, chest infections severe and frequent, consideration must be given to lobectomy before transplantation. If bronchiectasis is generalised the decision as to whether the patient can be accepted for transplantation rests on the frequency and severity of exacerbations of their infection.

Peptic ulceration When high dose corticosteroids were the mainstay of immunosuppression and before routine use of H_2 receptor blocking drugs (e.g. cimetidine and ranitidine), perforation of duodenal or gastric ulcers was a small but significant cause of mortality. It is now rare due to three approaches. Firstly, patients at risk can be identified on the basis of history and liberal use of endoscopic investigation in those with a positive history or relevant symptoms. Secondly, the high dose corticosteroid regimens are much less frequently employed. Finally, H_2 receptor blockade is used prophylactically in those at risk and in those who develop symptoms after transplantation, with special attention given during and soon after use of intravenous corticosteroid boluses.

Acute infection Almost all of the decisions needed to proceed to renal transplantation may be taken weeks, months or years before the operation. One detailed assessment can only be made on the day of operation and that is the possibility of current infection. Pyrexia is usually the best and most accurate guide since most patients will minimise or ignore their symptoms in a desire to be transplanted. Physical examination should concentrate on the upper and lower respiratory tracts; the abdomen and CAPD catheter exit site in those using peritoneal dialysis; and on the urinary tract. Routine samples of urine and swab of throat, nose, catheter exit site and skin flora in axillae and groins may help to guide antibiotic therapy after transplantation, especially if multiple resistance *Staphylococcus aureus* are found.

Long term risks

Cancer   Immunosuppression in patients with malignant neoplasia usually demonstrates the relevance of immune surveillance of neoplasia by rapid and aggressive metastasis. It is thus important to exclude malignant disease in all patients. In those known to have had tumours removed, it is prudent to wait a minimum of one year before contemplating transplantation, depending upon the expected natural history of the particular tumour and its initial staging. In some patients whose risk of recurrence is high, the neoplasm may act as a permanent contraindication to transplantation.

Liver disease   Chronic liver disease presents a number of problems. The first consideration is that drug handling, especially of cyclosporin, will be different to normal. The second problem is that many chronic liver diseases have been seen to progress quite rapidly, especially when azathioprine has been used. The third issue is that of chronic liver disease in patients with hepatitis B. There are data showing a poor prognosis in those with pre-existing hepatitis B due to liver failure after renal transplantation, especially when there is evidence of chronic liver disease before transplantation. Staff involved with care of such patients must be offered protective vaccination, but opinions differ on whether or not transplantation should be offered to patients with hepatitis B. There is, as yet, little data upon which to decide not to transplant a patient with chronic liver disease associated with hepatitis C and at present therefore these patients are offered renal transplantation.

Chronic infections   Infection with human immunodeficiency virus (HIV), whether or not it is associated with AIDS, represents an absolute contra-indication to transplantation because of the rapidly fatal course of the disease after active immunosuppression. Patients in the early 1980s, unfortunate enough to receive an HIV positive kidney died within months, of overwhelming infection, while those infected before transplant also suffered a dramatic downhill course.
   Chronic parasitic infections with malaria, schistosomes or other largely tropical organisms have not had a major impact upon transplant units, because the costs of transplantation have largely prevented those at greatest risk of such chronic infection from being considered. With more widespread use of renal dialysis and transplantation these issues will increasingly have to be faced.
   In principle, all chronic infections such as aspergillosis have to be fully treated before transplantation and effective prophylactic measures employed during the initial six month period of greatest immunosuppression.

Blood transfusion

Historical perspective

During the 1960s, as renal transplantation became an increasingly common technique for management of endstage renal failure, the presence of cytotoxic antibodies was recognised as a risk factor for hyperacute graft rejection (Chapter 5). It was known that blood transfusion, representing a mechanism of exposing a patient to foreign HLA antigens, was one way of stimulating cytotoxic antibodies. The assumption was therefore that blood transfusion was likely to be deleterious for a subsequent transplant. This posed many problems at a time when just to haemodialyse a patient might require transfusion of several units of blood. It was of interest to note, in retrospect, that there were three clinical observations and much animal data which suggested that blood transfusion was not in fact a risk factor for graft loss. In 1972, Opelz, Terasaki and their colleagues focused attention on data which showed that patients transfused before their renal transplant actually did significantly better than those who were untransfused. A sceptical profession spent the next five years studying the phenomenon in clinical practice and, having accepted the benefits, studying methods of dissecting those benefits from the disadvantages. It was thus not until the early 1980s that there was widespread acceptance of routine pretransplant blood transfusion. In 1986, data from both the Scandinavian transplant group and the Collaborative Transplant Study created by Opelz, were presented. These data showed quite convincingly that the one year graft survival of untransfused patients had been rising steadily, with the effect of abolishing the advantage of blood transfusion for patients transplanted after 1984. Many suggestions have been made as to the cause of these observed changes. The first to be proven incorrect was that cyclosporin might have led to the improved results, since the phenomenon was also observed in those treated with azathioprine and prednisolone only. If one is to rule out the suggestion that there has been a fundamental change in our immunobiology over a ten year time frame, then one is left with the suggestion that better diagnosis and management of acute rejection have prevented graft loss in untransfused patients, who would previously have rejected their grafts. There has thus been a return to scepticism about the advantages of blood transfusion. In view of these data most units now pursue a selective policy with 'routine pretransplant blood transfusion' of only those patients likely to be at least risk of developing cytotoxic antibodies. Table 2.5 details the untransfused and transfused patients' graft survival from some of the most influential studies during the evolution of blood transfusion practice.

Blood preparation

Many patients presenting for assessment prior to joining a renal transplant waiting list will have been transfused with blood at some time in the past. This blood will usually either have been 'whole blood' or 'packed cells', and will have been given at an operative procedure or during the course of managing their renal failure. Two questions arise in such patients: when

Table 2.5   The effect of blood transfusion before renal transplantation on graft survival

Study	1 year graft survival	
	Untransfused	Transfused
Opelz *et al.*	29	43–66*
Transplant Proc		
1973; **5**, 253–259		
Opelz & Terasaki	42	59–68
New Engl J Med,		
1978; **299**, 799–803		
Williams *et al.*	34	85
Lancet 1980; **1**, 1104–1106		
Sanfillippo *et al.*	46	55–61
Transplantation		
1984; **37**, 350–356		
Lundgren *et al.*	70	69
Lancet 1986; **2**, 66–69		
Opelz	81	83
Transplant Proc		
1987; **19**, 149–152		

*Range for different groups of transfused patients

were the transfusions; and how many units were given? A vague and unreliable history is, of course, the norm. The solution to this problem is now to presume transfusion, when in the past detailed and diligent searches of medical record archives would have been appropriate. If, on the other hand, the patient denies transfusion, what preparation if any should be given? In Table 2.6 a variety of alternative blood preparations have been compared. In order to provide a simple and effective blood preparation, most units would opt for plasma reduced blood ('packed cells') less than ten days old. If the indication for a transfusion is clinical necessity in a patient who has previously been transfused or who is at risk of developing cytotoxic antibodies, then the least immunogenic preparation is blood deprived of leucocytes by use of one of the current range of efficient blood filters.

How many units should one transfuse to observe an improvement in graft survival? It was much easier to answer this question in the 1970s than it is today, because of our doubts about whether the effect is still of clinical significance. The answer in the 1970s was that the more units the better, until one reached 20 units. Even then, the effect of 30 units might have

Table 2.6   Blood preparation for transfusion before renal transplantation

Preparation	Advantages	Disadvantages
Whole blood/packed cells	Availability	Sensitisation to HLA
Washed red cells	Reduced sensitisation to HLA	Ineffective for 'transfusion effect' on graft survival. Still has the potential to sensitise
Filtered blood	Low risk of sensitisation to HLA	Probably ineffective for 'transfusion effect'
Frozen blood	Low risk of sensitisation	Ineffective for 'transfusion effect'. Expensive processing
Buffy coat	Low volume	Sensitisation to HLA. Expensive processing

been greater than 20, but disguised by the poor overall prognoses of patients who needed that much blood. In practice, however, the majority of the observable effect could be achieved with three to five units of blood given on discrete occasions. Those centres who use routine pretransplant blood transfusion generally give only two or three units.

A representative protocol for transfusion of blood in patients awaiting renal transplantation is shown in Table 2.7. Despite all the doubt that remains over blood transfusion and the clinical consequences in transplant recipients, one factor has not altered. After a blood transfusion about 80% of patients who will develop cytotoxic anti-HLA antibodies will have done so within two weeks. All who become sensitised will do so within four weeks of the transfusion. It is therefore important to use serum samples taken a minimum of two and preferably four weeks after the transfusion, for crossmatching against potential donors (see Chapter 5). If one does not do this, there is a risk that the sample used for the crossmatch test will not be representative of circulating anti-HLA antibodies present at the time of the transplant.

Donor specific transfusion

With the exception of a single trial conducted in Germany, transfusion of blood from the kidney donor to the recipient is restricted to living donors. The suggestion was made on the basis of sound experimental evidence, that a blood transfusion from the kidney donor would greatly improve the outcome of that graft. The evidence from clinical practice has borne out this suggestion, with improved graft survival after donor specific transfusion for one haplotype mismatched living related donors (see Chapter 5). It is also interesting that the trial alluded to above, suggested that there could be a small advantage even with cadaveric donors. The major disadvantage of donor specific transfusion (DST) was a 30% incidence of donor specific antibody production, precluding the transplant. Indeed it has been suggested that the improvement in graft survival occurred because antibody production after DST specifically precluded from transplantation those grafts that would have been rejected. To a certain extent this has been disproved by simultaneous DST and immunosuppression with azathioprine

Table 2.7　Protocol for pre-transplant blood transfusion

Exclusions:	Parous females
	Males and nulliparous females with pre-existing anti-HLA antibody
Protocol:	Transfuse 1st unit of packed cells
	Two weeks later
	Test for cytotoxic antibodies, if present discontinue, if not:
	Transfuse 2nd unit of packed cells
	Two weeks later
	Test for cytotoxic antibodies, if present discontinue, if not:
	Transfuse 3rd unit of packed cells
	Wait four weeks
	Serum sample for transplant crossmatch

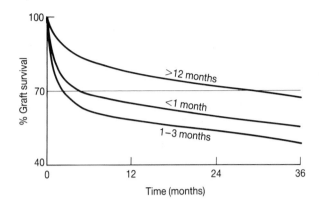

Figure 2.1 Renal transplant graft survival of second transplants stratified by duration, in months, of survival of the first graft. ($n = 382$, >12 months; $n = 206$, <1 month; $n = 114$, 1–3 months) (Collaborative Transplant Study, 1992).

or cyclosporin, where the incidence of donor specific antibody production has been reduced to 10% without abolishing the effect on graft survival. Data from the Collaborative Transplant study were influential in attitudes to DST, when it was shown that there was no difference in outcome between living related transplants transfused with blood from random third party donors and those given DST. Some centres have assumed that, by extension, there is now no difference between transfused and untransfused living related transplants, though this has not been shown conclusively.

Repeat transplantation

The decision to accept patients for a second transplant may be straightforward in the case of an early graft loss from arterial thrombosis. Neither the surgical suitability nor subsequent graft survival will have been significantly compromised and so the decision rests entirely with the patient's view of trying for a second time.

In most patients considered for a second, third or fourth transplant either their medical or surgical suitability will have changed as a result of the failed graft, or the reduced chances of a successful transplant (Figure 2.1 and discussed in Chapter 5) will alter the balance. Surgical suitability will rest upon the remaining options for vascular anastomoses and whether or not the failed graft has been removed (Chapter 11). Medical suitability is usually an issue only in those patients who have lost a graft after many years of function. In such patients, the only course is to assess them anew, rather than assume that their co-existing diseases, especially cardiac disease, have not progressed.

Further reading

1.   Cameron JS. Glomerulonephritis in renal transplants. *Transplantation* 1982; **34**, 237–245.
2.   Ost L, Landgren G, Groth CG. Renal transplantation in the older patient. In: *Progress in Transplantation 2.* PJ Morris, NL Tilney (eds). Edinburgh: Churchill Livingstone, 1985.
3.   Mathew TH. Recurrent disease after renal transplantation. *Transplant Rev,* 1991; **5**, 31–45.
4.   Opelz G. Improved kidney graft survival in non-transplant recipients. *Transplant Proc* 1987; **19**, 149–52.
5.   Marwick TH, Steinmuller DR, Underwood DA *et al.* Ineffectiveness of dipyridamole spect thallium imaging as a screening technique for coronary artery disease with endstage renal failure. *Transplantation* 1990; **49**, 100–103.

3   The cadaver donor

For almost every renal transplant programme, the supply of donor kidneys is insufficient to meet the increasing demand for transplantation. Living related and unrelated donors provide a proportion of transplanted kidneys, varying by necessity from 100% in some communities, to more commonly reported figures of 5 to 20% in most programmes. The balance of this demand must therefore be met by cadaver kidney donation, and has turned the thoughts of transplant units, health administrators and governments to tackle the complex medical, legal, cultural and ethical issues thought to hinder cadaver organ donation. Whereas most communities prefer to educate both the public and the medical profession in an attempt to close the gap between the number of actual and potential donors, others prefer to legislate for presumption that cadaver organ donation should take place.

Demand and supply

In North America, Europe and Australia, the demand for renal transplantation has been estimated in various ways to be between 50 and 70 patients per million population (pmp) each year. This estimate includes demand for retransplantation, which explains why it is the same as the number of patients accepted each year for dialysis, even though many patients suitable for dialysis are not transplant candidates.

In 1990, 17% of the 39 kidneys pmp transplanted in the USA came from living donors. In Australia, the dialysis and living donor rates are similar to that of the USA, but because of a comparatively poor cadaver kidney donor rate of 24 pmp, only 16.9% of the dialysis population were transplanted in 1990. The average time on the waiting list in Australia is three years. The donor rates of most developed countries also fail to meet the needs; however, the problems confronting many developing countries are very much greater. India, for example, with an estimated population of 893 million in 1990, has an estimated endstage renal disease rate of 90 to 100 pmp, almost no dialysis facilities and a renal transplantation rate of only 2.2 pmp. In 1990, 60% of the 2000 transplanted population of India received a kidney from an unrelated living donor, 40% from a related donor and 0.1% from cadaveric sources (0.02 pmp).

Having ascertained that up to 50 cadaver donor kidneys pmp are needed in most communities each year, one must ask if this number of cadaver donors suitable for renal donation is available. Evidence from selected parts of the world suggests that the potential exists. Firstly, there are individual regions and countries such as Austria and Norway that achieve this number of cadaver donors. Secondly, analyses restricted to deaths in intensive care units suggest that there is the potential to increase donor rates. Thirdly, detailed analysis of all deaths occurring in hospitals in Sydney, Australia, a city of 4 million residents, has shown that about 2% of deaths in hospitals are medically suitable potential organ donors. Hence, it can be estimated for most developed countries that there exist up to 50 cadaver donors (or 100 cadaver kidneys) pmp per year if all potential donors are identified, managed appropriately and if consent is granted.

Brainstem death

Central to the use and promotion of cadaver organ donors is an understanding and acceptance of the concept of brainstem death, a concept that permits retrieval of organs either in the presence of a beating donor heart or immediately after cessation of cardiac function. The advantages are a practical diagnosis of death and the provision of optimal physiological conditions for retrieval of viable cadaver kidneys for subsequent transplantation up to 48 hours later.

Definition of death

The traditional dictionary definition of a cadaver is one of a pale dead human body such as that preserved for anatomical study, which is clearly in a condition unsuitable for organ donation for subsequent transplantation. Alternatively, death recognised as cardiac standstill and observed as an irretrievable state by common experience has had fleeting moments in the past when its definition as cessation of the heart beat, could have been questioned. For example, when the guillotine fell to sever the head from the neck, none doubted that death had occurred despite continuing cardiac function that ensured the exsanguination of the beheaded body via the divided carotid arteries. Death and absence of cardiac function were not equitable then and are even less so now.

The difference, of course, is the capacity of current resuscitative medical technology to maintain biological function of the heart for long periods of time, through provision of artificial ventilation, intravenous electrolyte and fluid replacement and inotropic support of the myocardium. Diagnosis of an irretrievable state of brain death has thus been forced upon society, firstly in order to prevent prolongation of pointless, distressing and expensive treatment, and secondly, to facilitate the use of cadaver organs for transplantation. The definition of death is now based on irretrievable loss of the capacity for consciousness, to breathe and to sustain spontaneous cardiac function. It can therefore be defined in terms of brainstem function, with the reticular activating system responsible for consciousness and the respiratory centre for breathing.

Historical perspective

Shakespeare provided both Hamlet who gave voice to thoughts of death, and Romeo, whose inadequate diagnosis of Juliet's brain death confirms for some people their fears of this new definition of death. Similar doubts have necessitated determination of unequivocal criteria to make a simple and widely applicable diagnosis of brain death, which can be understood easily by both the medical profession and non-medical people.

The concept of brain death was first discussed in 1959 and many publications outlining medical and subsequent legal criteria for brain death followed. The two most important analyses were the 'Harvard Criteria for Brain Death' in 1968 and the 'Report of the Conference of the Royal Colleges and Their Faculties' of the United Kingdom in 1976. The former, setting the groundwork upon which subsequent guidelines were based, required an absence of responses to painful stimuli, absence of spontaneous movement, absence of reflexes and a flat electroencephalogram. The importance of the later Royal Colleges' Conference criteria, also known as the 'UK Code' and which has received wide acceptance, was that the diagnosis of brain death could safely be based entirely on clinical criteria and would not require investigation. These criteria continue to be refined by clinical experience and experimental research and most countries have since legislated that irretrievable loss of brainstem function constitutes death.

Aetiology of brainstem death

The brain is a pliant structure situated within the fixed volume of the bony confines of cranial cavity and is susceptible to raised intracranial pressure and resultant reduction in cerebral blood flow. The brainstem, the critical region of the brain responsible for maintaining respiration and activation of the cerebral cortex above, is a relatively fixed structure that traverses the rigid hiatus in the fibrous tentorium which separates the cerebral hemispheres above from the cerebellum in the posterior fossa below (Figure 3.1).

Catastrophic events such as massive trauma to the head, rupture of the

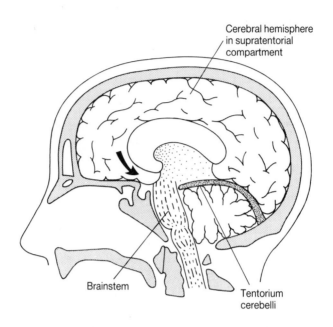

Cerebral hemisphere
in supratentorial
compartment

Brainstem

Tentorium
cerebelli

Figure 3.1 Schematic representation of the cranial cavity and its contents with the tentorium cerebelli separating the supratentorial from the infratentorial compartment. The arrow demonstrates the downward direction of the movement of supratentorial cerebral tissue to compress the brainstem.

brainstem by judicial hanging, basilar artery thrombosis or abrupt cessation of blood flow in both carotid and vertebral arteries are direct infratentorial or primary causes of irreversible brainstem damage and death. Without ventilatory support in such instances, the heart will continue to beat for periods of up to 20 minutes and until such time as systemic hypoxia produced by the sudden respiratory arrest causes myocardial arrest.

Brainstem death secondary to the volume effect of supratentorial pathology, is, however, more common. Massive pressure increases within the supratentorial compartment forces the undersurface of each cerebral hemisphere through the tentorial hiatus. In doing so, the brainstem is damaged by the combination of direct compression, interruption of the arterial supply and impairment of venous drainage. This process of herniation of cerebral tissue, commonly described as 'coning', can follow extradural, subdural and intracerebral bleeding, and cerebral swelling associated with contusion or the effect of anoxia on brain cells following a period of circulatory arrest. A seemingly irreversible chain of events is set in motion by a severe supratentorial injury which is in turn compounded by progressive reduction of cerebral blood flow and consequent cerebral hypoxia and swelling (Figure 3.2). The time between the onset of the injury and brainstem death following 'coning' often depends on the severity of initial injury and may take minutes, hours or days. Eventually, the raised intracranial pressure prevents cerebral blood flow and the consequent irreversible cessation of all functions of the whole brain.

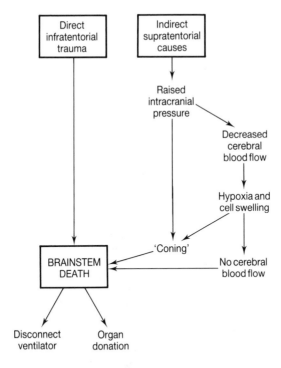

Figure 3.2 Sequence of events leading to primary (direct) or secondary (indirect) brainstem death.

The causes of brainstem death and the numbers of cadaver donors will vary according to the locality of the donor centre. The National Organ Donor Registry of Australia in 1990 reported that 42% of organ donor brain death resulted from cerebral haemorrhage, 46% from trauma involving road traffic accidents, falls and gun shot wounds, while 12% were from other causes including asthma, cerebral tumours and drownings. Legislation enforcing the wearing of safety belts in motor vehicles and stringent policing of drink driving laws have reduced the number of trauma related donors and placed increased emphasis on donors with cerebral haemorrhage. Figure 3.3 demonstrates the biphasic distribution of the age and sex of donors at time of death. The principal cause of death in the peak of younger and especially male donors was predominantly due to trauma, while the second peak in the fifth decade was associated with cerebral haemorrhage and a female preponderance.

Preconditions to a clinical diagnosis of brainstem death

Many sets of roughly similar guidelines exist for the clinical diagnosis of brainstem death and it is important that criteria defined by local legislation are used. When followed in a stepwise fashion, these guidelines are a safeguard against mistakes. The first and most important step is to satisfy all the preconditions summarised in Table 3.1. Diagnosis of the responsible

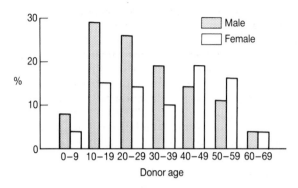

Figure 3.3   Age and sex distribution of cadaver organ donors in Australia in 1989/90.

structural condition for the coma is usually apparent with a history of a severe head injury, a subarachnoid haemorrhage or prolonged hypoxia, and can be supported by cerebral CT scan findings. The decision that the causative pathology is irreversible can only be made after all feasible therapeutic measures, such as evacuation of an intracranial haematoma and correction of hypotension, have been made to reverse the effects of the pathology.

The necessary passage of time between presentation of a patient in a state of coma and formal testing for brainstem death can vary according to circumstances. Its major value is in allowing time for appropriate therapeutic measures to be undertaken and to exclude functional causes or components of the comatose state. Whereas four to six hours may be sufficient after a severe head injury or a subarachnoid bleed from a known cause, for other potential donors, 24 hours or longer may be necessary to reverse hypothermia (temperature <35°C) or gross metabolic disturbances. Days may be necessary to exclude possible drug intoxication if screening facilities are not available.

Clinical testing

The purpose of satisfying the preconditions is to ensure that the clinical tests for brainstem death are performed on appropriate patients at the right time. The aim of the testing, which is comparatively straightforward, is to

Table 3.1   Clinical preconditions for the diagnosis of brainstem death

1.   Comatosed patient on a ventilator

2.   Known cause of irreversible structural brain damage

3.   Exclusion of functional causes of coma:
     * hypothermia
     * neuromuscular blockade
     * gross metabolic disturbance
     * alcohol intoxication
     * central nervous system depressant drugs

4.   Passage of time to undertake therapeutic measures and satisfy the preconditions

establish that the patient has absent brainstem reflexes and is apnoeic. The number of medical practitioners needed to make the diagnosis of brainstem death, and their mandatory level of qualification, varies from country to country, but generally includes formal testing at the bedside by two practitioners with more than five years clinical experience. In practice, these practitioners are usually intensivists, anaesthetists, neurologists or neurosurgeons. Medical ethics and brainstem death legislations stipulate that those involved in establishing the diagnosis of brainstem death must not have interests in, or care for, patients who may benefit from transplantation of cadaver donor organs. At the time of completion of tests confirming brainstem death, the patient can be declared legally dead. The attending clinician may then discontinue ventilatory support or, if appropriate, seek consent to retrieve organs for subsequent transplantation.

Conveniently, the brainstem contains cranial nerve nuclei responsible for reflexes that can be tested in the comatose patient, with each reflex involving a motor response to a sensory input (Table 3.2). Spinal reflexes may still be present and do not exclude the diagnosis of brainstem death. Whereas fixed dilated pupils are a variable finding in a brainstem dead patient, the absence of a pupillary response to a bright torch light in a darkened room is a consistent finding. The examiner must ensure that mydriatic eye drops have not been used in the previous four hours. Corneal reflex testing is performed using firm pressure with a cotton tipped swab rather than the conventional light touch employed in the awake patient. The vestibulo-ocular reflex involves installation of 20 ml of ice cold water into the external auditory canal, with failure of eye movement of either eye during an observation period of one minute indicating an absent reflex. The presence of otorrhoea or a ruptured eardrum prevents this test from being performed. Testing of the gag and cough reflexes requires temporary disconnection from the ventilator, although experienced nursing staff, whilst performing routine suction of the oropharynx and lower airway, often note the progressive loss of responses during the evolution of brainstem death. If there is any doubt about the results of brainstem reflex testing, they should be repeated after an appropriate interval.

Table 3.2   Clinical testing for absence of brainstem reflexes

Brainstem reflexes	Sensory cranial nerve	Motor cranial nerve
1. Absent pupillary response to light	II	III (parasympathetic fibres)
2. Absent corneal reflexes	V	VII
3. Absent vestibulo-ocular reflex	VIII	III, VI
4. Absent cranial nerve response to pain	V	VII (and limb motor responses)
5. Absent gag and cough reflexes	X	IX

In order to determine respiratory centre death, it is necessary to demonstrate that spontaneous respiration does not occur in the presence of the stimulatory effect of a full carbon dioxide drive. Hypoxia is avoided by ventilation of the patient with 100% oxygen for five minutes before disconnection from the ventilator. During the test period, 100% oxygen is provided through a tracheal catheter. In the brainstem dead patient, carbon dioxide tension increases at the rate of 0.3 kPa/min or 2 mmHg/min during apnoea. If the initial carbon dioxide tension is about 5.3 kPa (40 mmHg), the arterial carbon dioxide tension will be in the vicinity of 8 kPa (60 mmHg) approximately ten minutes after disconnection of the ventilator. This should be sufficient stimulatory drive to produce spontaneous respiration if the respiratory centre is intact. It is necessary to confirm the presence of high carbon dioxide tension by measurement of arterial blood gases. The test may be difficult to undertake in patients with severe chest trauma causing hypoxaemia, or those with chronic obstructive airways disease who are dependent on hypoxia to stimulate respiration.

Diagnosis by investigation

Clinical tests may be impossible to perform in cases of facial trauma, or

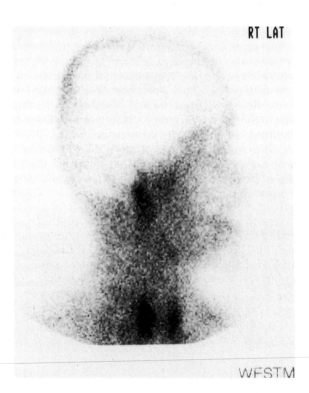

Figure 3.4 ^{99m}Tc-HMPAO radionuclide brain scan demonstrating absent cerebral blood flow in a brain dead patient. The rim of the blood flow that surrounds the cranial cavity is supplying the scalp.

cannot be performed in the presence of paralysing or sedating drugs such as barbiturates. In such instances, investigation of whole brain function can be used to complement clinical testing of the brainstem. Electro-encephalography, cerebral angiography, radionuclide brain scanning and transcranial Doppler ultrasound measurements have been used to demon-strate lack of supratentorial brain function or absence of cerebral blood flow (Figure 3.4). All these techniques have limitations; however, if angiography demonstrates the absence of cerebral blood flow, then brainstem death can be confirmed. If angiography is inconclusive, the circumstances can be explained to the relatives and the patient may be taken to the operating suite and prepared for donor surgery. The patient is then disconnected from the ventilator and the donor procedure commenced after observation of cardiac standstill. In this circumstance, the conventional definition of death has been fulfilled, in that irreversible cessation of cardiac function has been demonstrated.

Despite confirmatory investigations, there remains uncertainty about the reliability of clinical brainstem testing in young children, particularly neonates. It is recommended that organs for transplantation should not be removed within the first seven days of life from neonates with beating hearts. For children under the age of one year, repeat clinical testing and confirmatory radionuclide brain scanning have been suggested.

Criteria for cadaver kidney donation

A variety of medical conditions preclude organ donation entirely, or may exclude use of particular organs. Careful consideration of potential donors with respect to medical suitability is important before discussing organ donation with the brainstem dead patient's relatives. It is unwise to raise the potential for organ donation in a situation where medical suitability has not been addressed. The two principal considerations of medical suitability are firstly, whether the kidney will function satisfactorily after transplantation and secondly, whether transplantation of the kidney or other organs will transmit disease. Details of the absolute and relative contraindications to donation of a kidney are listed in Table 3.3.

Around 70% of cadaver donors become multi-organ donors with each transplant speciality having their own criteria which depend on the age and size of the donor, and parameters of specific organ function. The disparity between demand and supply has forced transplant units continually to reassess and widen their criteria to accept more marginal donors in order to reduce waiting lists. How then is it possible for intensive care units to assess medical suitability? The simple answer is to consider organ donation in all brainstem dead patients and to refer to the local organ retrieval team for a decision on medical suitability.

Quality of donor kidneys

The likelihood of chronic renal disease receives more attention than parameters of acute dysfunction after transplantation when assessing the quality of a donor kidney. Acute renal failure or dysfunction in the donor certainly implies similar problems are likely in the recipient, but many kidneys have functioned well in the long term despite the need for temporary dialysis. Kidneys from hypertensive and diabetic patients can be accepted for transplantation provided baseline renal function is normal and macroscopic appearance is not grossly abnormal. If in doubt, a needle core biopsy taken at the time of kidney retrieval can be urgently processed prior to subsequent transplantation.

The currently accepted age range for cadaver kidney donation is 0 to 70 years although there are studies that indicate reduced success of transplantation when the donor age is less than five or greater than 60 years (see Chapter 11, Figure 11.2). The glomerular filtration rate is estimated to decrease by 1 ml per minute for every year of age above 40. Hence, donor age will correlate with graft function and it can be argued that kidneys from aged donors should be transplanted to older recipients. For the very young donor where there is a risk of arterial thrombosis in the small renal arteries, consideration should be given to retrieving and transplanting kidneys 'en bloc'. There appears to be no advantage in transplanting kidneys from donors less than five years of age into recipients less than five years old.

Medical exclusion criteria

Because of the risk of transmitting malignant cells, potential donors with a past history of malignancy other than non-melanoma skin tumours or primary brain tumours should be excluded. In addition, the potential for later development of renal malignancy excludes potential donors with Hippel-von Lindau syndrome. All donors are also tested for hepatitis B and hepatitis C which, if present, exclude organ donation except perhaps for the purpose of transplantation into recipients already infected with the relevant virus. Cytomegalovirus infection is not regarded as a contraindication to transplantation, though some centres will ensure that positive donor kidneys are only transplanted into recipients who have had previous infection.

Particular attention must be paid to risk factors for transmission of HIV infection. A history of drug abuse, prostitution or male homosexuality may

Table 3.3   Contraindications to cadaver kidney donation

Absolute
1.   Respiratory/circulatory arrest for more than 30 minutes
2.   Malignancy other than primary brain and non-melanotic skin tumours
3.   HIV infection
4.   Male homosexuals/bisexuals, intravenous drugs users and their sexual partners
5.   Hepatitis B and C infection
6.   Chronic impairment of renal function

Relative
1.   Acute renal dysfunction
2.   Age greater than 70 years
3.   Age less than three years
4.   Hypertension
5.   Diabetes
6.   Bacteraemia with known organism and sensitivities

---

have been disguised from the next of kin and be available only from close friends. It is often difficult to raise the issue of such high risk behaviours, but it is of major importance. Presence of any previous medical and social history, including being a sexual partner to a person in this high risk group, should be considered as a very strong contraindication to organ donation. Though all donors will be tested for antibody to HIV, it is acknowledged that patients who have been infected within the previous three months will be infectious and are unlikely to have developed antibody.

Consent for organ retrieval

Consent to proceed with organ donation is, in most countries, the next major barrier. The process of gaining formal consent from the relatives or from the patient during life can be defined as 'opting in'. Unless consent is expressly given, the presumption is that consent is withheld. In a number of countries such as Belgium and Singapore, the opposite pertains. Consent is presumed unless the patient has specifically 'opted out' before death. This undoubtedly influences organ donation rates favourably, but may be seen as intrusive on individual rights.

Refusal of consent by relatives occurs in up to 50% of those asked in North America, Europe and Australia. Reasons for refusal include the inability of relatives to accept a sudden and unexpected death; the intrusive nature of the request; perceived mutilation of the body; and deeply ingrained religious convictions or cultural beliefs that are not amenable to change. Nevertheless, long experience from intensive care units that have many organ donors has shown that it is impossible to predict the reactions of the family. Some relatives may prefer to consent to donation of some organs and not others.

The greater the experience of the person asking for consent, whether they be an intensivist, neurosurgeon, transplant coordinator or social worker, the greater the chance that consent will be given. It is unwise to discuss brain death and its consequences with a patient's family without being able to answer their questions on organ donation. Furthermore, many clinicians experience discomfort in approaching relatives and discussing the concept of brainstem death, and any perceived awkwardness on their part may influence relatives adversely.

Reluctance to seek consent is understandable but wrong, for it is the right of the relatives to be given the opportunity to consider cadaver organ donation. Some governments and hospital authorities have attempted to bypass this reluctance by enforcing 'required request' or 'routine enquiry'. It may, however, be more productive for all parties involved to work together to overcome the pyschological and other barriers and provide a working environment in which cadaver organ donation is accepted as part of the continuing medical management of the brainstem dead patient.

Logistics of organ retrieval

Organisation of a donor procedure can be a complicated task and on

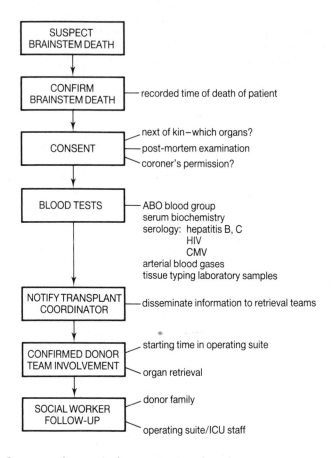

Figure 3.5   Sequence of events in the organisation of a cadaver organ donor.

occasions can in itself be seen by referring intensive care units to be a disincentive to cadaver organ donation. A well defined unit protocol helps, and if not already involved, this is the time to contact the transplant coordinator (Figure 3.5). Their major role, after consent has been given, is to bring together, as rapidly as possible and at any hour of the day, the appropriate surgical teams for possible retrieval of kidneys, heart, lungs, liver, pancreas, small bowel, eye and bone. This may involve liaison with different hospitals, cities, states, countries, or even continents.

The norm would be to expect six to 12 hours to elapse between the time of diagnosis of brainstem death and cross clamping of the aorta, and for the multi-organ donor procedure to take three to five hours to complete. These times may be clearly inappropriate in the haemodynamically unstable donor, which calls for more rapid responses and restriction of the number of possible organs retrieved. Single donor teams capable of retrieving multiple organs make for easier, less disruptive and more cost effective donor procedures, particularly for smaller hospitals.

It is essential that the costs of any tests ordered after declaration of brainstem death, necessary to provide information for the donor teams, as well as the cost of maintaining the donor in the intensive care unit and operating theatre, are not billed to the donor in private health care systems. A postmortem examination is performed after the donor procedure.

Donor management

The completion of the diagnosis of brainstem death and obtaining consent for organ donation heralds a change in management priorities, for no longer is it necessary to direct treatment towards brain survival, but rather towards providing optimal conditions for subsequent survival of the organs after transplantation. Even with maximal physiologic support, 80% of potential donors will suffer cardiac arrest within 120 hours. The dominant problem is hypotension, either as a result of hypovolaemia from previous aggressive diuretic therapy or as a consequence of loss of anti-diuretic hormone (ADH) release leading to diabetes insipidus which occurs in at least one third of donors.

The principal objective of donor management, both before and during the donor procedure, is maintenance of a stable circulatory system with a systolic blood pressure greater than 90 mmHg and a urine output of at least 1 ml/kg/hour. This is best achieved with intravenous crystalloid fluid replacement and avoidance, where possible, of inotropic support. Diabetes insipidus can be diagnosed by low urine and high serum osmolality, and treatment should include replacement of urinary fluid and electrolyte losses and use of synthetic ADH preparations to reduce urine output to between 2 and 5 μg/kg/hour.

A number of attempts have been made to manipulate the recipient's immune response to the kidney by pretreating either the donor or the perfused organ. Blood transfusion of the donor, treatment with cortico-steroids and high dose cyclophosphamide have all had positive results in single trials. Removal of donor bone marrow and transplantation into the recipient at the same time as the kidney, in an attempt to create a mixed chimeric state in the recipient, is currently under trial. Determined attempts to alter the immunogenicity of the donor kidney have used lytic monoclonal antibodies targeted at the dendritic cells. The fact that none of these protocols is in widespread or routine use may testify to their current efficacy.

Surgical techniques

The donor surgical procedure is performed in an operating suite with an anaesthetist, but unlike any other operation, the patient is already dead and the procedure ends in transfer of the body to the mortuary. This difference may lead to profound effects upon the operating suite staff and must be recognised as placing an emotional burden upon all who are involved, no matter how often. The quality of the donor surgical procedure is as important as transplant procedure and should not be the task of untrained surgeons. The donor surgeon verifies that the diagnosis of brainstem death has been made according to law and that appropriate permissions have been gained for organ donation.

'Kidneys only' donation

The aim of the donor surgeon is to remove the two kidneys in an unhurried manner, with minimal handling and care to preserve the kidney vasculature. A wide bilateral subcostal incision is made and a thorough laparotomy undertaken. The retroperitoneal structures are displayed by mobilising the small bowel mesentery and right hemicolon. The right and left crus, superior mesenteric artery and coeliac trunk are divided. The inferior vena cava is mobilised above and below the renal veins, the distal aorta mobilised and an aortic cannula inserted. Isolated perfusion of the kidneys is further facilitated by division of the inferior mesenteric artery, gonadal and lumbar vessels. A separate cannula is placed in the inferior vena cava through which venous drainage of the perfusion fluid is achieved. After systemic heparinisation with 20,000 units of sodium heparin, the inferior vena cava and aorta are clamped above and below the renal vessels and cool perfusion commenced simultaneous with discontinuation of ventilation (Figure 3.6). At least two litres of cold perfusion fluid are passed into the aortic cannula at a pressure of 75 to 100 mmHg.

The retroperitoneal dissection of the kidneys, together with ureters down to and beyond the level of the pelvic brim is then completed. The kidneys can either be separated *in situ* or removed 'en bloc' and separated on the

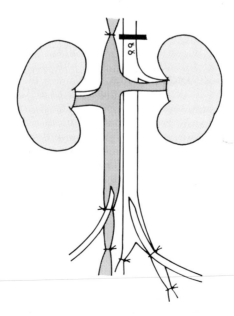

Figure 3.6   *In situ* cold perfusion of the isolated segment of cross clamped abdominal aorta and drainage of venous effluent from the inferior vena cava.

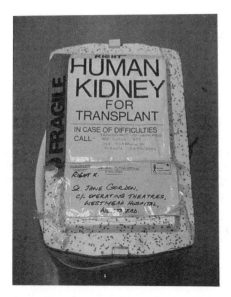

Figure 3.7 Cadaver donor kidney packed in an ice storage box for transportation to transplanting centre.

back table. Each kidney is individually packed into three sterile plastic bags inside each other, labelled as either left or right, and then packaged in ice ready for distribution after matching the donor to the most appropriate recipient (Figure 3.7).

Multi-organ donation

It is usual for the liver retrieval team to attend the operating suite first. Simultaneous retrieval of the heart, lungs, liver, pancreas and kidneys is facilitated by a median sternotomy and an abdominal incision continued down to the pubic symphysis. After a laparotomy, the small bowel mesentery and right hemicolon are mobilised to display the retroperitoneal structures including the aorta, inferior vena cava and kidneys. Structures of the porta hepatis are identified and the liver and pancreas are mobilised with preservation of the coeliac trunk and superior mesenteric artery in preparation for subsequent 'en bloc' removal of the two organs. The distal aorta, inferior vena cava and superior vein are then mobilised in preparation for insertion of perfusion cannuli.

The heart retrieval team are then invited to the operating table for the comparatively short mobilisation of the heart and, if appropriate, 'en bloc' with the lungs. Cold aortic and portal perfusion are commenced simultaneously with introduction of cold cardioplegic solution to the ascending aorta and cross clamping of the proximal abdominal aorta. At this stage the anaesthetist disconnects the ventilator and ceases cardiac monitoring. Between two and four litres of cold preservation perfusion fluid are infused into the distal aorta and the portal vein, cooling the liver, pancreas and

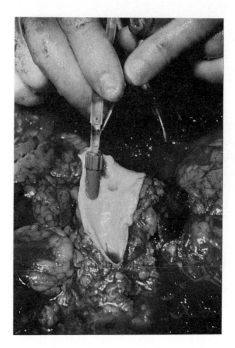

Figure 3.8   *Ex situ* perfusion of the right renal artery. Both kidneys have been removed 'en bloc'.

kidneys. The kidneys are cooled at the rate of 10°C per minute when the perfusion fluid is at 4°C. The heart, liver and pancreas are removed, followed finally by retroperitoneal dissection of the kidneys.

Non-heartbeating donor

Legislation in some countries, sudden death of a haemodynamically unstable donor, or the use of executed prisoners as organ donors, necessitate the use of a very different surgical technique for organ retrieval. Before the widespread acceptance of brainstem death in the late 1970s, the situation was the norm for clinical transplantation, with the prolonged warm ischaemia time associated with non-heartbeating organ donation usually precluding retrieval of the heart, lungs and liver.

Kidneys can tolerate up to 30 minutes of warm ischaemia time before becoming non-viable. If cardiac standstill occurs in the operating suite, it is thus possible for experienced retrieval surgeons to remove two kidneys en bloc and cold perfuse on the back table (Figure 3.8). An alternative technique is to insert a specially designed aortic perfusion cannula with two balloons capable of occluding the aorta above and below the renal arteries. Once cooled, the kidneys can be removed in a more controlled manner.

Organ preservation

The effectiveness of kidney preservation is measured by the quality of early kidney function after transplantation. Factors affecting that quality of preservation include donor management, nephrectomy technique, warm ischaemia time, total renal ischaemia time and the anastomosis time. The length of time that a kidney can be stored depends on both cooling to diminish metabolic activity and thus oxygen requirements, and on use of fluids designed to preserve the intracellular *milieu* in the absence of a sodium/potassium pump. Since surface cooling is too slow and freezing would produce irreversible intracellular damage, perfusion of preservation fluid at 2°C to 4°C is undertaken through the renal arteries.

Perfusion with normal saline would result in transmemembrane influx of sodium, loss of potassium and rapid destruction of cellular integrity. A number of fluids have been designed which retain cellular viability for many hours or days. Cellular oedema is reduced by use of either a slowly permeant or impermeant solute such as mannitol, sucrose or raffinose. Intracellular pH is preserved with buffers such as phosphate, citrate or lactobionate, and high concentrations of potassium prevent influx of sodium and water. Commonly used solutions are either isotonic and mimic the intracellular electrolyte composition (Collins, Marshall, Eurocollins), or hypertonic such as Ross citrate solution. Any of these solutions preserve cold stored kidneys for up to 24 or 48 hours.

The University of Wisconsin (UW) solution is more complex, containing adenosine and glucose as a metabolite supply, allopurinol and glutathione as free radical scavengers and metabolic inhibitors, and insulin and steroids as membrane stabilisers. Initially developed for cold storage of the pancreas and used extensively for liver preservation, the solution provides no clear advantage over other preservation solutions for kidneys stored for up to 24 hours, but may be more effective thereafter.

An alternative to cold storage of the perfused kidneys packed in ice, is continuous machine perfusion. In essence, a pump is used to recirculate either oxygenated synthetic stable albumin based solution or UW solution at temperatures between 5°C and 10°C. Good renal preservation for up to 72 hours has been shown, but continuous perfusion has the disadvantage of being labour intensive and is unnecessary if the kidneys are transplanted within 24 hours of donation.

The donor family and bereavement

Sudden and unexpected death of a family member is always devastating for those left behind to grieve. Brainstem death provides a further problem for the process of bereavement because the family must take on trust that the warm, pink, ventilated body that they said goodbye to in the hospital intensive care ward was indeed dead. Even when the body has been laid to rest, the doubts and uncertainties remain.

Privacy and time are two commodities in short supply in most intensive care units, yet it is provision of those two factors at the time of diagnosis of brainstem death which may help the family in subsequent weeks, months and years. Development of trust in the staff caring for the patient and understanding the concepts of brain death may allay future nightmares. Those involved in intensive care units and with transplantation need to develop an understanding of bereavement, look for those families in need of help and then provide professional counselling, support and guidance. To ignore this problem not only detracts considerably from the benefits of transplantation, but also provides a steady trickle of unhappy and disturbed families antagonistic to transplantation and ready to generate negative publicity.

Strategies to increase cadaver organ donation

Despite apparent social consensus accepting organ donation and required request laws in some states and countries, the conversion of potential to actual organ donor remains at about 20%. This would suggest that there are unresolved problems both within the intensive care units that seek consent for organ donation, and with the public who fail to give consent.

Directed towards intensive care units

The attitudes of staff caring for patients who are already, or may become, potential organ donors are a major determining influence on cadaveric transplantation. The legal requirement to ask for consent from suitable donors' families was clearly seen as a mechanism for increasing the rate of organ donation. Unfortunately, the anticipated improvements did not always occur and in a number of states of the USA, the donor rates actually dropped. This result is not surprising if uninterested and untrained hospital staff are designated to get a form filled out in which they sign that consent for organ donation has been requested.

Training, experience and skill are more likely to increase donor rates. Patients who suffer from major traumatic head injury or subarachnoid haemorrhage and obviously have a hopeless prognosis may be treated by ventilation, inotrope and fluid support until formal assessment of brain death, or careful judgement of prognosis and no further resuscitation. The latter course is much easier to take, and may be viewed as the most appropriate clinical management of the patient. The attitudes that determine which of these courses of action is taken are those of intensivists, neurosurgeons, accident and emergency teams, and the nursing staff on the appropriate wards. Transplantation is promoted as being full of glamour and healthy happy patients, whereas intensive care units see the opposite face with dead patients, disturbed families and unfolding tragedies. Until these two disparate views can be brought together effectively, utilising some of the benefits to offset some of the tragedy, organ donation rates seem likely to remain at levels that do not meet the demand.

Directed towards the public

Renal transplantation has been a clinical reality for only 30 years and it is therefore not surprising that surveys have demonstrated a paucity of knowledge in the community about organ donation and transplantation. In comparison, religious, cultural and ethnic attitudes to death have developed over centuries and to change those attitudes will no doubt take longer than 30 years. Even though many religious laws and leaders condone organ donation (Table 3.4) many communities and countries do not accept cadaver organ donation.

There are several ways of changing the attitude of the community towards organ donation. One is by multimedia advertising campaigns which often use well known public figures to provoke discussion. Recent campaigns in Australia have failed to produce an instant increase in the organ donation rate. This may in part be due to costs limiting the extent of the multimedia coverage and to the fact that the issues pertaining to the complexities of organ donation are difficult to address in a 20 second television advertisement.

An alternative but long term strategy is to create an informed society of young adults who understand and appreciate all the issues involved in transplantation. A cross curriculum programme for secondary schools has been developed in Australia with an education package involving comprehensive teaching notes and a video containing case studies. The students are encouraged to discuss these issues in the family environment and it is expected that this programme will profoundly influence community attitudes and awareness about organ donation and transplantation in the years to come (Figure 3.9).

Table 3.4 Religious attitudes to organ donation

Religion	Attitude to organ donation
Amish	Reluctant
Baha'i	Yes
Buddhist	Individual choice
Christian Scientist	Individual choice
Episcopal Church	Yes
Greek Orthodox	Yes
Gypsies	No
Hinduism	Yes
Islam	Yes
Jehovah's Witness	Not encouraged
Judaism	Yes
Mormon	Yes
Protestantism	Yes
Quakers	Yes
Roman Catholic	Yes

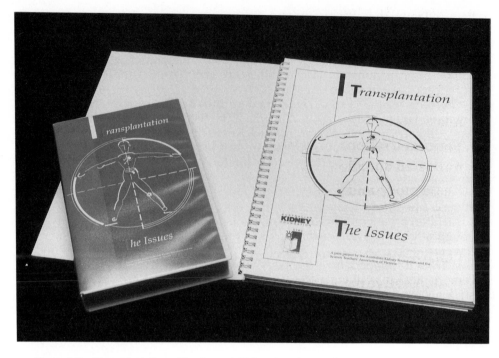

Figure 3.9 *'Transplantation: The Issues'*. This education package is available from STAV Publishing P/L, PO Box 190, Richmond, Australia, 3121.

Further reading

1. Pallis C. Brainstem death. In: *Handbook of Clinical Neurology, Volume 13(57): Head injury.* Braakman R (ed). Amsterdam: Elsevier Science Publishers, 1990.
2. Sommerauer JF. Brain death determination in children and the anencephalic donor. *Clin Transplant* 1991; **5**, 137–145.
3. Belzer FO, Southard JH. Principles of solid organ preservation by cold storage. *Transplantation*, 1988; **45**, 673–676.
4. Brannigan MC. A chronicle of organ transplant progress in Japan. *Transplant Int*, 1992; **5**, 180–186.

4 The living donor

The first successful renal transplant was a living related transplant between two identical twin brothers performed in Boston in 1954 and reported in 1956. That transplant not only saved the life of the donor's brother, since there was no satisfactory alternative treatment, but also demonstrated that renal transplantation was a viable option for endstage renal disease.

With development of the alternatives of haemodialysis, peritoneal dialysis and cadaveric renal transplantation, living donation is no longer, at least in the developed world, the only alternative to death. In the less developed world and where there are cultural and religious barriers to cadaveric organ donation, living donation still remains the single alternative to death for many recipients. In practice living donation from a related donor constitutes at least 10% of all transplants performed worldwide in the 1990s. The operations and postoperative care of living donors have been refined over the years to ensure a progressively safer procedure, though there have been a small number of recorded cases of death of the donor after renal donation. The gap between success rates after living donation and cadaveric donation have narrowed over the years, acting as further disincentive to living donation. No matter what these competing influences might have achieved in theory, while cadaveric donor organ supply fails to meet demand, the pressure for living donation will surely continue unabated.

Ethics and rationale for living related donation

Many ethical issues have been debated with respect to the use of living related donors for renal transplantation. Any person involved in living organ donation has first to take stock of his or her ethical standpoint. Living organ donation is almost unique in the field of medicine in that the procedure performed on a patient is not for their benefit and has physical disadvantages. One can argue that the psychological and non-specific benefits to the donor are real, particularly when a close relative is returned to normal health. There can, however, be no doubt that the physical consequences of living donation are entirely detrimental to the donor. It is relatively easy to understand the motives behind the first living kidney donation and one may assume that living donation between relatives carries the same altruistic motives. One may therefore accept that donation between relatives implies that the donor receives a temporary physical disadvantage and exposes him or herself to a temporary, small, but real risk, while gaining in the long term the knowledge that their relative has returned to normal health. Experience stands in the way of this simple creed. While many donors fall neatly into this altruistic categorisation there are, unfortunately, many examples where the donors have exacted physical, emotional or financial toll from the recipient from the day of donation.

For the recipient, there are two specific advantages which may be conferred by living related donation. The first is that transplantation can be planned and timed to coincide with the appropriate point in the patient's decline of renal function. Patients can be transplanted before the requirement for dialysis treatment and at a time that suits their symptoms and social needs. As a consequence, living related donation will also take place during the daytime in a fully staffed operating theatre with return of the recipient to a ward where the staff are expecting their arrival. It is not clear whether there are material benefits from these factors, since attempts to measure them have been sketchy. It is known, however, that transplanting patients before their need for dialysis does not confer specific immunological advantage or disadvantage, given current immunosuppressive regimens. The second advantage, as discussed below, comes from the improved results of renal transplantation from living related donors.

Ethics and rationale for living unrelated donation

There are a number of different issues that have to be analysed when the putative living donor is unrelated to the recipient. One must consider the unrelated donor who has a stable and close emotional relationship with the recipient, such as a husband or wife. It is clearly possible in this situation to draw an analogy between living related donation and this form of unrelated donation. While it is usually possible to determine that financial inducement is not a factor, the possibility of coercion should always be considered. If one believes that living unrelated donation in this instance is reasonable, the next step is to consider the living unrelated donor who has no personal relationship with the recipient. There can be such unrelated donors where the motivation is clearly not financial. The first and most difficult is the family donor unsuitable for reasons of blood group or lymphocytotoxic crossmatch for donation to a particular recipient. Some have suggested that, in return for living donation to the most appropriate person selected from a pool awaiting cadaveric donation, the relative of the donor would receive priority for an appropriate cadaveric kidney from the pool. The second situation is where a potential donor decides that their most effective contribution to society would be to donate a kidney, unconcerned as to whom it is given. While one can write off the latter case as a person who is almost by definition psychiatrically disturbed, it is much harder to dismiss the first case as totally irrelevant.

Having addressed and conquered these issues, one arrives at the ethics of paid unrelated organ donation. Whatever the perceptions of this practice in developed countries, it is widespread across the world and while specifically addressed by The Transplantation Society's statement on the ethics of transplantation, by the World Health Organisation and by law in many countries, paid unrelated donation is happening.

The rationale for unrelated and unmatched donation is much harder to define in the same scientific terms as for matched related donation. There is certainly no clear advantage to unmatched donation in terms of graft or patient survival when one compares unmatched living with cadaveric donation. There is, however, a difference in the initial function rate, with living donors not surprisingly providing the most reliable initial function. The rationale for unmatched donation has perhaps to be specific to the alternatives available to the patient. If neither dialysis nor cadaveric donation are available, the alternative is clearly no survival, compared with perhaps 70 or 80% at one year.

Each person involved in transplantation must arrive at a conclusion with which they are comfortable, given their clinical practice and the ethical and cultural environment in which they work.

Blood group and tissue type matching living donors

With the exception of isolated practices, living donation has been performed between ABO blood group compatible donors. The major advantage conferred by living donation relates to the degree of histocompatibility matching between the donor and recipient as discussed in Chapter 5. Matching of the antigens of the major histocompatiblity complex reduces the risks of rejection. In Figure 4.1 the three different degrees of match are shown: completely matched, half matched, and completely mismatched. The tissue type in almost all families is inherited en bloc, half from the patient's mother and half from the father. Thus, by definition, the patient will share one group of tissue types, known as a haplotype, with his or her mother and the other group with his or her father. Similarly, the patient's children will have inherited one of the patient's haplotypes, while receiving the other from the patient's spouse. When searching for a donor from amongst the patient's siblings, there are three alternatives. Firstly, the sibling may have inherited neither of the same haplotypes from the parents and therefore be tissue type unmatched. Secondly, the sibling may have inherited either the same haplotype from the mother as the patient or the same haplotype

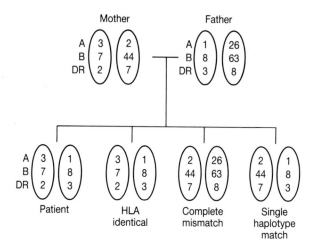

Figure 4.1 Inheritance of HLA tissue types. Diagrammatic representation of the alternative combination of HLA tissue type in a family of a patient with 3 siblings. Each haplotype is inherited independently, one from the father and one from the mother. Siblings 1 and 2 are identical while sibling 3 is entirely mismatched. Sibling 4 has inherited the same haplotype as the patient from their father but different haplotypes from their mother and is thus a one haplotype match.

from the father, but received the unmatched haplotype from the other parent. These siblings are thus single haplotype matches and are therefore equivalent to the patient's parents, children or quarter of the cousins. The final situation, for which there is a one in four chance, is inheritance of both of the same haplotypes. Potential donors may therefore be fully matched (two haplotypes), half matched (a haplotype match) or unmatched.

The results of two haplotype matches and single haplotype matches demonstrate considerable advantage for living related over cadaveric transplantation, the completely matched living related donor providing approximately a 97% chance of successful transplantation at the end of one year, with continuing advantage accruing year by year thereafter. In data from the Collaborative Transplant Study, the calculated half-life of an HLA identical sibling transplant is 23.6 years (Figure 4.2). The single haplotype match donor provided a 90% chance of success at one year and reaches 52% after ten years (half-life = 11.9 years). In contrast, cadaveric transplantation (without regard to HLA matching) conferred an 82% one year graft survival, a half-life of 8.8 years and a ten year survival estimate of 40%. One large renal centre (Seattle in the United States) has performed a careful analysis of the factors which affect patient survival, as opposed to graft survival. In their analysis, having tried to account for the differences in the patients who are selected for treatment by transplantation and dialysis, no clear advantage could be seen for cadaveric transplantation over dialysis, but there was a significant prolongation of the patients' lives when living related transplantation was undertaken.

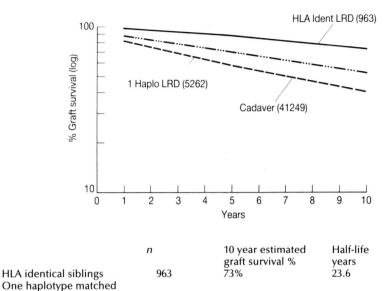

	n	10 year estimated graft survival %	Half-life years
HLA identical siblings	963	73%	23.6
One haplotype matched			
Living donors	5262	52%	11.9
Cadaver donors	41249	40%	8.8

Figure 4.2   Outcome of transplantation by source of donor. All patients received first transplants and were treated with cyclosporin (data provided by courtesy of the Collaborative Transplant Study).

Assessment of the living donor

General approach

Considerable care is needed to avoid disrupting and destroying families with the stresses of living donation. Each potential living donor should be approached on the basis that their donation may fail and that the patient to whom they donate may lose their life. While some donors are determined to donate, many are not and they and their families may find themselves in an emotional trap where the donor feels pressured to donate, but is actually ambivalent or antagonistic to the idea. It is thus important to ensure that all living donors are assessed predominantly in privacy, and that while the initial discussion about organ donation and transplantation may involve the recipient, no subsequent part of the assessment should do so. Any donor who wishes to withdraw his or her offer of organ donation and for whom it thus becomes medically unsuitable to proceed can then explain to the potential recipient that they have withdrawn because of medical reasons or their change of heart, since this is a decision for the donor and not the doctor.

In the 1960s and 1970s, it was common practice for all potential donors to be assessed for both medical suitability and bona fide intentions by an independent physician. In addition, consultation with a psychiatrist was also undertaken to verify the voluntary and informed nature of the decision taken by the donor. While this practice still continues in some institutions, others believe that a sensitive and informed approach first by a renal physician and then by a transplant surgeon provides a fully acceptable alternative.

History

In almost all countries with formal legislation governing living related donation, the donor must be over the age of legal consent. When there is any doubt about the potential donor's eligibility it is therefore reasonable to seek proof of age. An assessment must be made of the potential donor's level of intellectual capacity, since neither an accurate history nor informed consent would be possible from a severely impaired person.

The potential donor's formal medical history, and in particular the potential donor's previous medical history, is of critical importance to exclude known hypertension and renal or other disease, such as diabetes, which may threaten the donor's own renal function in the future. Presence of any chronic ill health is almost certain to exclude the potential donor from being a satisfactorily low risk for anaesthetic. Assessment of current medications and known allergies permits identification of risk factors, such as oral contraceptive use, which need to be identified and eliminated before donation. Current smoking or alcohol abuse needs to be excluded, or lead to exclusion of the potential donor. Apart from the specific medical history, it is important to gain an assessment of the patient's social history including pastimes, occupation and plans for a family. One must feel uneasy about recommending removal of one kidney from people who indulge in dangerous physical occupations (such as active service in the armed forces) or indulge in pastimes where trauma, including loss of one kidney, is a real risk (such as contact sports or motor cycle racing). Removal of a kidney from a young woman about to embark upon pregnancy is less than desirable.

Many renal diseases can be inherited in forms that cause renal failure in the recipient but may not be overt in a younger sibling. If the recipient has reflux nephropathy, care must be taken to be sure that the potential donor does not also suffer the same problem. Inherited diseases, like polycystic kidney disease and Allport's syndrome, would alert one to particular lines of examination and investigation. Similarly, the possibility of recurrence of the recipient's original disease in the transplanted kidney must be taken into consideration and the potential donor informed of the possibility (see Chapters 2, 10, 11).

Examination

Full routine physical examination of the potential donor, by both renal physician and surgeon, in an unhurried atmosphere will be needed to exclude physical signs of disease. Clear concise recording of the details and measurement of the blood pressure on more than one occasion is required as a standard.

Table 4.1   Assessment of a blood group compatible living donor

1.	*History*
	Relationship to recipient
	Age
	General history
	Medical history
	Medications (oral contraceptive)
	Allergies
	Social history – occupation, hobbies, sports, HIV transmission risk factors
	Family history
	Motivation
	Alcohol, tobacco drug abuse
2.	*Examination*
	Full routine physical examination
	Repeat BP on two or more occasions
3.	*Investigation*
	Check blood group compatibility
	Tissue type, T and B cell crossmatch,

Blood
  – urea, creatinine, electrolytes,
  – uric acid, liver function tests, Ca, PO_4
  – glucose,
  – fasting cholesterol/triglycerides,
  – full blood count, ESR, clotting indices
  – viral studies, CMV, HIV, hepatitis B, hepatitis C, EBV
  – VDRL
  – GFR estimation (Cr-EDTA, or Tc-DTPA or creatinine clearance)

Urine
  – midstream urine culture, microscopy (repeat)
  – urinalysis for protein, glucose, haematuria, leucocytes
  – 24 hr urine protein

Chest X-ray, IVP, ECG
Selective renal angiogram
Repeat lymphocytoxic crossmatch within one week of transplantation

4. *Particular issues relating to recipient disease*

Reflux nephropathy	– repeat midstream urine
	– micturating cystogram may be needed
Polycystic kidney disease	– renal ultrasound, and proceed if donor age >30yrs
Diabetes mellitus	– glucose tolerance test
Allport's syndrome	– audiometry, urine microscopy for haematuria
Haemoglobinopathy	– haemoglobin electrophoresis

Investigation

Investigation of potential living donors has three aims. Firstly, the donor must be fully fit for anaesthesia, secondly the donor must have two normal kidneys and thirdly, the donor must not have any signs of the disease causing renal failure in the recipient. It is clearly appropriate to determine blood group and tissue type prior to proceeding to the complete evaluation of the potential donor, as outlined in Table 4.1. In investigation of the potential donor, it is wise to recall the related recipient's disease, since there are a number of investigations that are determined by the need to uncover subclinical or asymptomatic disease in donors. There are a number of tests which can be undertaken as an outpatient without significant risk to the donor. Of these, only the intravenous pyelogram carries the potential of direct physical hazard (the risk of anaphylaxis being estimated at 1:100,000). The intravenous pyelogram can be replaced by DTPA scanning at the time of estimation of the glomerular filtration rate, providing that attention is subsequently given to the renal anatomy outlined during the angiographic assessment. Testing for human immunodeficiency virus carries implications for follow-up and counselling of any potential donors found to be positive. Providing that these tests are satisfactory, it is necessary to proceed to evaluation of the renal vasculature by aortogram. A number of centres have used peripheral venous injection and digital subtraction angiography before organ donation, but clear definition of the presence or absence of small accessory renal arteries can be difficult using this procedure and most centres have therefore reverted to conventional aortograms. Providing that no contraindications have emerged during the donor's assessment (Table 4.2), the final test before transplantation is to repeat the lymphocytotoxic crossmatch test in the week before transplantation (see Chapter 5).

Living donor nephrectomy

The surgical team involved in performing a living donor nephrectomy have the most onerous task in clinical transplantation. The donor is, in every respect, a healthy individual and expects to leave hospital in a similar state of health but with the addition of a surgical scar. Furthermore, the ease of the recipient's operation and early transplant function is dependent on intact renal anatomy and the quality of renal preservation. For these reasons, living donor nephrectomy must be performed by an experienced and senior surgeon.

Table 4.2   Contraindications to living kidney donation

Relative
1.   Smoking
2.   Obesity
3.   Hyperlipidaemia
4.   Hypertension
5.   Accessory renal arteries
6.   Duplicated ureteric drainage

Absolute
1.   Renal disease
2.   Abnormal anatomy such as solitary kidney, horseshoe kidney
3.   Underlying renal disease, e.g. polycystic kidneys, glomerulonephritis
4.   Medical conditions, e.g. insulin dependent diabetes mellitus, coronary artery disease, neoplastic disease (other than localised skin cancer), infection with hepatitis B, hepatitis C, human immunodeficiency virus
5.   Intellectual impairment
6.   Coercion or financial reward

Pre-operative assessment

In addition to the medical assessment detailed above, the transplanting surgeons must, themselves, ascertain the relationship between the donor and recipient. This is especially important in countries such as the United Kingdom where unrelated donation is illegal, except in special circumstances, and only then after formal approval has been given. The minimum evidence needed is tissue typing data from the family, together with taking a family history from both donor and recipient, perhaps including such details as the mother's maiden name. Similarly, proof of the donor's age may be appropriate if there is doubt that they are over the age of legal consent. Obtaining consent necessitates a description of the operation and the possible complications. Pain associated with the large incision and avoidance of heavy manual labour for six to 12 weeks after the surgery, are both inevitable consequences (see Table 4.3). It is not possible to obtain informed consent from a potential donor who is intellectually impaired.

Anatomical considerations

When there is a choice, donation of the left kidney is preferred because the longer renal vein on that side is easier both to remove and to transplant. Review of the angiogram and intravenous pyelogram may, however, dictate otherwise. A kidney with two or more arteries, or with duplicated ureters, is more difficult, though not necessarily impossible, to transplant. Kidneys with such anatomical variations are more likely to be lost from technical misadventure.

Operative technique

After induction of anaesthesia and muscle relaxation, a small gauge urinary catheter is inserted to drain intra-operative diuresis. A single prophylactic injection of a broad spectrum antibiotic and measures to reduce the risks of deep vein thrombosis are usual. The donor is then placed in a lateral position on an operating table flexed by 10°. This opens the space between the costal margin and the iliac crest to facilitate surgery. For cosmetic reasons, there is a temptation to limit the size of the incision, but particularly in muscular

Table 4.3   Potential complication of living donor nephrectomy

---

Wound
Infection, haematoma, hernia

Respiratory
Atelectasis, pneumonia, pneumothorax, haemothorax, pleural effusion, pulmonary embolus

Urinary
Acute tubular necrosis, urine infection, urine retention

Other
Ileus, deep vein thrombosis

Late complications
Hypertension, proteinuria, wound discomfort

---

or obese donors, this may compromise the function of the transplanted kidney because of direct injury or excessive handling needed during the nephrectomy. On the left side, the incision can be made from either the space between the 11th and 12th ribs, or through the bed of the 12th rib, obliquely downwards. On the right side, the kidney is lower, permitting a flank incision without rib resection. If there are multiple arteries to the donor kidney, an anterior transperitoneal approach may be advisable.

The lateral incision involves division of the latissimus dorsi and the three muscle layers of the abdominal wall in order to enter the retroperitoneal space. Care must be taken to avoid damaging the pleura posteriorly, since this may precipitate a pneumothorax and require insertion of a chest drain. The fascia surrounding the kidney is divided and the kidney mobilised, preserving the adrenal gland and its blood supply, the perihilar fat and, most importantly, the tissue between the lower pole of the kidney and the ureter. Excessive handling and traction on the renal vessels will cause vasospasm and should thus be avoided. The ureter is mobilised with care to preserve the blood supply and divided, after ligation, at the pelvic brim. It is important to obtain the maximum length of the renal vessels by division flush with the aorta and inferior vena cava. Following ligation and division of the vessels, the kidney is handed to a separate perfusion team who cannulate the renal artery and perfuse the kidney with 300–500 ml of cool (4°C) organ preservation fluid, at a pressure of 75–100 mmHg. After closure of the wound and during recovery from the anaesthesia a routine chest X-ray is needed to exclude pneumothorax.

Postoperative management

In most circumstances, the postoperative management is best undertaken in a ward separate from the transplant recipient, away from the sights and sounds of management of the recipient's postoperative course. Management of the living donor following nephrectomy is the same as for a conventional nephrectomy. Even when the retroperitoneal approach has been used to provide access to the kidney, paralytic ileus is common. Care is therefore required in resumption of oral fluid intake, but neither nasogastric tubes nor urinary catheters are required as a routine. The operation, involving rib resection and division of muscle layers, is painful. The donor will require much more analgesia than the recipient, given either by epidural, intravenous or subcutaneous opiate infusion. Because of the proximity of the wound to the rib cage, attention to chest physiotherapy is simultaneously the most important and usually the most active intervention required. Mobilisation of patients takes up to five days with hospital discharge within one week. Throughout this period it is advisable to use prophylaxis for deep vein thrombosis. Most donors are able to return to work within three to six weeks depending upon the nature of the work they undertake, but up to 12 weeks may be needed to return to unrestricted physical activity. The potential complications that may be encountered are detailed in Table 4.3, though the approximate incidence of complications that should be quoted to patients will depend, to a certain extent, upon local experience. Wound infection may occur in approximately 2% of cases, while chest complications including atelectasis (10–15%) and pneumothorax (10%) are the most common. Urinary tract infection, usually related to catheterisation of the bladder, is sufficiently common (5–10%) to recommend midstream urine culture prior to discharge. Mortality from the operation is the most difficult factor to estimate and advise upon accurately. While there are known to have been a number of deaths amongst living related donors, the true incidence of mortality is not known. On the basis of undocumented reports a figure in the region of 1 in 4000 may be correct, with death from pulmonary embolism being the commonest cause.

Long term management

It is common practice to assess all living donors six weeks and six months after the donation, though living renal donors seldom require long term management. Many centres have, in the past, regarded it as necessary to monitor donors in the long term, in order to assess the potential long term complications of such operations. It is from these follow-up data that a 10–15% incidence of late hypertension has been noted. This incidence of hypertension is similar to that of hypertension in relatives of patients with renal disease and may thus have little to do with the nephrectomy. Nevertheless, the ability of the glomerular filtration rate of the kidney to respond to a protein load is suboptimal in living related kidney donors. It has not, however, been demonstrated that this has any long term implications for renal function and life insurance companies do not routinely load their premiums after nephrectomy.

While physical complications are those which cause physicians and surgeons most concern, it is the psychological impact of kidney donation which may have the most significant long term outcome. It is therefore important to pay particular attention to those donors where renal donation has failed to result in a successful transplant.

Further reading

1. Anderson RG, Bueschen AJ, Lloyd LK, Dubrovsky EV, Burns JR. Short and long term changes in renal function after donor nephrectomy. *J Urol* 1991; **145**, 11–13.
2. Higashihara E, Horie S, Takeuchi T, Nutahara K, Aso Y. Long term consequences of nephrectomy. *J Urol* 1990; **143**, 230–243.
3. Council of the Transplantation Society. *Lancet* 1985; **2**, 715–716.

5 Matching donor and recipient

The first task of the transplant unit with a waiting list of patients in need of renal transplantation is to establish a mechanism for matching the donor kidneys with suitable recipients, as they become available. This implies that the unit has decided the criteria that will be used to allocate kidneys and the order of priority of those criteria. There are criteria used by almost all units, such as blood group compatibility. Even here there is scope for differences in practice, since some units may offer blood group O kidneys to all patients, while others will restrict the offers to blood group O recipients. Other criteria vary greatly between units, such as discrimination on the grounds of age or clinical urgency. The major issues in matching donors and recipients are addressed in this chapter.

Age

The age of donor and recipient has little or no known relevance until the extremes of age are reached. The age of the donor affects graft survival no matter what the age of the recipient. Kidneys from donors under the age of five years have about a 10% worse graft survival at one year than from adult donors, while those under two years are not used by many transplant units because the graft survival at one year is 30% worse. There are two main reasons why young kidneys do less well: firstly, the vascular and ureteric anastomoses are more technically demanding and ensuing complications lead to a greater incidence of graft loss, particularly from thrombosis; secondly, there appears to be a higher chance of losing a very small kidney to allograft rejection. One approach to the use of kidneys from donors under two years is to transplant both kidneys with the aorta and inferior vena cava, en bloc (see Chapter 15). At the other end of the age range, kidneys from donors over the age of 60 years also have a slightly lower graft survival rate than those from donors between five and 60 years. It is unclear whether this is due to the effect of age itself or to higher incidence of diseases such as hypertension in the elderly donors.

The suggestion that one should match donors for age, at the extremes, thus stems from the surgical practicalities of anastomosis and renal bulk in small donors and recipients and from the concept that an elderly kidney will not last for the young patient's anticipated life time. The evidence suggests that young recipients do better when given an adult kidney rather than a paediatric donor kidney.

Tissue typing

The earliest experiments of Gorer and Medawar in the 1930s and 1940s demonstrated the existence of 'histocompatibility antigens', matching for which determined the fate of transplanted tumours, skin and organs. The only renal transplants performed successfully without immunosuppression were those between identical twins. Attempts to transplant between non-identical individuals failed until the advent of azathioprine and cortico steroid immunosuppression in 1961. This stimulus to discover and match for tissue antigens thus lead Payne, van Rood and Dausset, amongst others, to use sera from patients who had received multiple blood transfusions or had multiple pregnancies, to define the different tissue antigens which determine graft rejection.

The major histocompatibility complex

The major histocompatibility complex (MHC) is a region of genetic material, on chromosome 6 in humans, which codes for the major antigens involved in allograft rejection. There are three classes of molecule encoded: class I, class II and class III (Figure 5.1). This genetic region has also proved to contain segments which encode many of the molecules which have, in recent years, been shown to be intimately involved in the immune response to foreign antigen and tumours. The immunological roles of class I, II and III molecules are discussed in Appendix 1.

Class I molecules   Human leucocyte antigens (HLA) -A, -B and -C are all of a similar molecular structure (Figure 5.2) with variable segments which define the exact specificity. HLA-A molecules from different individuals differ in their exact amino acid sequence in a few positions – sufficient to make the molecules distinct to the antibodies and T cell receptors with which they interact. Every person has two different HLA-A antigens, one on each chromosome, two HLA-B and two HLA-C. The currently recognised class I molecules are shown in Table 5.1.

Class II molecules   The class II molecules first became apparent because differences in class II antigens between two individuals cause lymphocytes

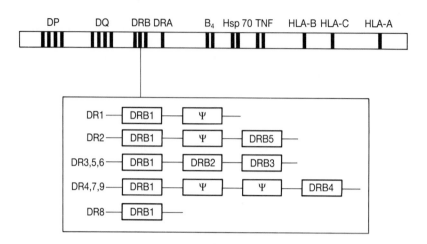

Figure 5.1   The major histocompatibility complex on the short arm of chromosome 6. Class II genes are represented by HLA-DP, DQ, DR; class I genes by HLA-A, B, C. The different gene arrangements for HLA-DR present in individuals with different HLA-DR types are shown inset in the box. Class III molecules are represented by B4, HSP 70 and TNF.

to proliferate in a mixed lymphocyte response (MLR). In the mid 1970s it became possible to distinguish some of the differences by using antibodies directed at these molecules, which are present largely on the surface of B lymphocytes. Thus the HLA-DR and more recently HLA-DQ tissue types were defined by serological techniques. Cellular and molecular genetic techniques have now taken definition of class II to the detail shown in Figure 5.1 and Table 5.1.

Inheritance   The MHC genetic region represented in Figure 5.1 is present in two copies in each individual. A complete chromosome is almost always

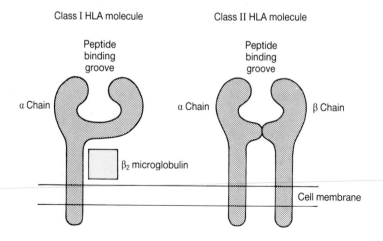

Figure 5.2   Diagrammatic structure of the HLA class I and class II molecules.

Table 5.1 HLA Nomenclature (WHO Nomenclature report 1991)

HLA-A	HL-B	HLA-B	HLA-C	HLA-D	HLA-DR	HLA-DQ	HLA-DP
A1	B5	B50(21)	Cw1	Dw1	DR1	DQ1	DPw1
A2	B7	B51(5)	Cw2	Dw2	DR103	DQ2	DPw2
A203	B703	B5102	Cw3	Dw3	DR2	DQ3	DPw3
A210	B8	B5103	Cw4	Dw4	DR3	DQ4	DPw4
A3	B12	B52(5)	Cw5	Dw5	DR4	DQ5(1)	DPw5
A9	B13	B53	Cw6	Dw6	DR5	DQ6(1)	DPw6
A10	B14	B54(22)	Cw7	Dw7	DR6	DQ7(3)	
A11	B15	B55(22)	Cw8	Dw8	DR7	DQ8(3)	
A19	B16	B56(22)	Cw9(w3)	Dw9	DR8	DQ9(3)	
A23(9)	B17	B57(17)	Cw10(w3)	Dw10	DR9		
A24(9)	B18	B58(17)		Dw11(w7)	DR10		
A2403	B21	B59		Dw12	DR11(5)		
A25(10)	B22	B60(40)		Dw13	DR12(5)		
A26(10)	B27	B61(40)		Dw14	DR13(6)		
A28	B35	B62(15)		Dw15	DR14(6)		
A29(19)	B37	B63(15)		Dw16	DR1403		
A30(19)	B38(16)	B64(14)		Dw17(w7)	DR1404		
A31(19)	B39(16)	B65(14)		Dw18(w6)	DR15(2)		
A32(19)	B3901	B67		Dw19(w6)	DR16(2)		
A33(19)	B3902	B70		Dw20	DR17(3)		
A34(10)	B40	B71(70)		Dw21	DR18(3)		
A36	B4005	B72(70)		Dw22			
A43	B41	B73		Dw23	DR51		
A66(10)	B42	B75(15)		Dw24	DR52		
A68(28)	B44(12)	B76(15)		Dw25	DR53		
A69(28)	B45(12)	B77(15)		Dw26			
A74(19)	B46	B7801					
	B47						
	B48	Bw4					
	B49(21)	Bw6					

passed from each parent to each child and with it the genetic definition of one HLA-A, one HLA-B and one HLA-DR specificity together with the intervening genes. This segment of gene is described as the haplotype and is usually transferred intact from one generation to the next. In a small proportion, crossover can occur with transfer of part of one haplotype and part of the other. Certain combinations of HLA-A, B and DR antigens occur more commonly than would be expected if there was a random distribution of antigens in the population (e.g. HLA-A1, B8, DR3 is the commonest haplotype in Caucasoids), a phenomenon termed 'linkage disequilibrium'. It is because of this fact that the seemingly impossible task of matching HLA antigens between unrelated donors and recipients becomes a partial reality.

Tissue type matching

Prospective, deliberate and exact matching of donors and recipients for the histocompatibility antigens was the hope of early research into tissue typing. This hope was shown to be increasingly unrealistic, except in the setting of a living related transplant, as the complexity of the 'tissue type' became apparent. Partial matching for HLA-A and -B antigens did, however confer a small advantage, in the short term, with graft survival for well matched grafts being 5–10% better. When HLA-DR typing became practical and of sufficient accuracy, studies quite quickly showed considerable advantage of matching for HLA-DR, -B and -A with a 20% improvement in survival at one year when all six antigens were matched. Calculation of match can be done in two ways – match or mismatch. Mismatch in fact describes the immunological situation most accurately and is thus used in most analyses. The stimulus to graft rejection is an antigen present on the kidney which is foreign to the recipient, i.e. the mismatch. Matching of antigens is less relevant and gives a different answer if the donor is homozygous for a particular antigen, i.e. has the same antigen on both chromosomes (Table 5.2).

The current results of both prospective (deliberate) and retrospective tissue matching show a particular advantage to matching for HLA-B and -DR, with less contribution from HLA-A. Data collected from around the world by both the Collaborative Transplant Study and the University of California Los Angeles (UCLA) Registry are shown, together with those collected by the Australian ANZDATA registry, in Table 5.3. It is thus accepted that tissue type matching, when it can be achieved, provides an advantage in recipients of first transplants. There is less general acceptance

Table 5.2   HLA matching versus mismatching

	HLA-A	HLA-B	HLA-DR	
Donor	1,24	27,35	3,3	
Recipient	2,24	7,8	3,4	
Matched antigens*	24		3	= 2 antigen match
Mismatched antigens**	2	27,35		= 3 antigen mismatch

*   A matched antigen is present in both donor and recipient
**   A mismatched antigen is present in the donor but not recipient

Table 5.3 Tissue type matching results for HLA-A, -B and -DR; 1st cadaver transplant recipients, 1 year graft survival

Mismatches	Collaborative Transplant Study %	UCLA Registry %	ANZDATA Registry %
0	86	85	93
1	84	85	88
2	81	81	85
3	80	80	81
4	79	76	78
5	75	73	79
6	74	73	93

that it should be used on a prospective basis as the sole method of allocating kidneys, as discussed below.

Minor histocompatibility antigens

That there are other antigens capable of eliciting an immune response when present on a transplanted organ or tissue, is not in doubt. The clearest evidence of their existence comes from clinical practice, where rejection of a sibling's HLA identical kidney or bone marrow (or the reverse direction of graft-versus-host disease) demonstrates that HLA identity is not sufficient to prevent an allograft response.

It has been possible, in both mice and man, to show that there are many genetic loci capable of initiating a minor histocompatibility response. The clue to the mechanism by which they do so, comes firstly from the fact that the response is confined to T cells and does not involve antibody production, and secondly that it is MHC restricted. MHC restriction implies that if one generates a T cell clone that is cytotoxic to cells which possess a particular minor histocompatibility antigen, then it will only kill cells that express not only the minor antigen but also a particular major histocompatibility antigen. It is from these experiments that we have come to believe that minor histocompatability antigens are in fact peptide fragments of normal cellular molecules that have been processed and bound into the antigen binding groove of either class I or class II HLA molecules.

While minor histocompatibility antigens have clinical significance, it is neither possible (since they are as yet not well enough defined) nor practical (since there are too many possible combinations) to match for these minor antigens. It may, in the future, prove possible to use information on the degree of minor antigen mismatching to predict the relative requirement for immunosuppression in particular patients. At present, however, there is no clinical strategy for exploiting knowledge of the minor histocompatibility antigens.

Blood group matching

In the report of the first series of renal transplants performed in Boston, it was assumed that incompatibility of blood group would probably be a factor to avoid when matching donors and recipients. It was concluded that ABO incompatibility might have been the explanation for rapid loss of one of the grafts. During the 1960s there was also an assumption that blood group compatibility was important, but despite this a number of ABO incompatible grafts were performed. When the results of these grafts were collected together it was clear that the presumption was correct, with only very rare grafts functioning in the long term. It is thus right to conclude that ABO compatibility is a requirement of donor-recipient matching. Detailed analysis of the Rhesus and Lewis systems, however, suggests that incompatibility of these blood groups can be ignored in routine practice.

There are two situations in which incompatible blood groups have not led to rapid graft rejection and perhaps explain the fact that about one third of the 92 ABO incompatible transplants reported to the Collaborative Transplant Study were still functioning one year after transplantation. The first is when the incompatible blood group is A_2, where it has been shown by the Scandinavians that graft survival may be close to the norm, in distinction to A_1 incompatibility. This might be explained by the decreased density of blood group A on the kidneys of A_2 compared to A_1 individuals.

The second series in which ABO incompatible transplants have been performed is where living related but ABO incompatible grafts have been transplanted into well prepared recipients. Preparation involves splenectomy and removal of naturally occurring antibodies by use of membranes which bind blood group molecules, together with immunosuppression. There are not many centres willing or equipped to undertake this degree of preparation and it is not surprising that ABO incompatible transplants represent less than 0.2% of the operations performed across the world.

Crossmatching

The basis of the crossmatch test in renal transplantation is the same as that used in blood crossmatching, in that it is designed to demonstrate antibody/antigen interaction. While agglutination is a successful approach to routine red cell crossmatching it was rejected as insensitive in leucocyte crossmatching.

The most commonly used test is the complement dependent lymphocytotoxicity assay, designed to demonstrate anti-HLA antibodies. There are now a variety of methods which augment the sensitivity of this test or enhance its specificity. There are also some centres using alternative tests which either use different technology to examine the same interaction (flow cytometry crossmatching) or examine antibodies specific for antigens carried on vascular endothelium, monocytes or epithelial cells.

The lymphocytotoxic crossmatch

Serum samples   In this test, samples from the recipient are placed into the wells of microtitre plates in 1 or 2 μl volumes. The serum tested always includes the patient's most recent sample, usually taken within the previous month, though some would insist on a sample taken on the day of the test. This is called the 'current sample' and is expected to reflect the antibody

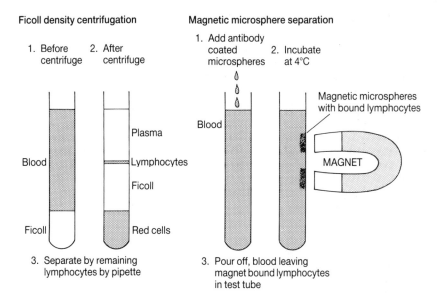

Figure 5.3 Principles of two alternative methods for separating peripheral blood lymphocytes: ficoll density centrifugation and magnetic microsphere separation.

status at the time of the planned transplant operation. Care must therefore be taken in interpretation of the 'current crossmatch' result if the serum is actually one or two months old. Any factor that might alter the patient's antibody status occurring after the sample had been taken, such as a blood transfusion, would render the test invalid. A second sample would be used in many patients, especially if they have previously been shown to have antibodies, by testing serum samples against a panel of randomly selected normal individuals (see below: Antibody screening). It would be conventional to use one or more of these historical sera taken at the peak of the patients' previous sensitisation. Interpretation of a positive crossmatch test with one of these peak sera varies from centre to centre as described below.

Donor Lymphocytes   A total of 2000 lymphocytes in 1 μl of medium is needed for each well of the microtitre plate. The source of donor lymphocytes may be peripheral blood, surgically excised lymph nodes or spleen. Peripheral blood lymphocytes (PBL) have proved to be unsatisfactory because of lack of sensitivity, when they are separated using the standard ficoll density separation technique (Figure 5.3). This has been presumed to relate to the physical state of the donor, together with the effects of medication, especially corticosteroids. Lymph node gives the most easily separated lymphocyte preparation, while the spleen, though satisfactory, requires further purification steps to remove macrophages. The alternative method of extracting pure lymphocyte preparations from blood, lymph node or spleen is to use magnetic beads coated with monoclonal antibodies of the appropriate specificity (Dynabeads®, Figure 5.3). Magnetic bead separated lymphocytes are more sensitive in the crossmatch test and so the test incubation times have to be reduced correspondingly. The other advantage with the magnetic bead technique is that peripheral blood samples can be used. The combination of peripheral blood as a source, reduced incubation times and prolonged surgery from multi-organ harvesting now permits allocation of kidneys on the basis of tissue match and crossmatch result, before nephrectomy has occurred in some instances.

Lymphocytes may be used without further purification into subsets, in which case PBL would contain approximately 90% T lymphocytes, 10% B lymphocytes, while lymph node and spleen have perhaps double the percentage of B lymphocytes. All B but only a few T lymphocytes express class II antigens and so antibodies to HLA class II antigens can only be detected reliably by separating B lymphocytes using nylon wool, sheep erythrocyte rosetting, or Dynabeads specific for class II molecules. All centres use the T cell crossmatch but many ignore the B cell crossmatch, uncertain about its relevance to transplant outcome.

Crossmatch method   Incubation of donor lymphocytes in recipient serum for a fixed period is followed by addition of rabbit complement for a further fixed period. The temperature at which these steps are incubated and the time periods vary from laboratory to laboratory and from one rabbit complement source to another. A selection of such times and temperatures is shown in Table 5.4.

The resultant antibody/antigen/complement interaction may be detected in a number of ways, each dependent upon the loss of cell membrane integrity due to formation of the membrane attack complex of complement. Eosin or Trypan Blue pass through the membrane of lysed cells but not of live cells, giving a different appearance through phase contrast light microscopy, of positive and negative crossmatch results. Using fluorescence techniques, necessary with dynabead separated lymphocytes, acridine orange and ethidium bromide combine to give orange fluorescent dead cells and green fluorescent live cells.

Antibody screening   It is both possible and useful to predict an individual patient's chances of having positive crossmatches with potential donors. Most centres will screen serum samples from patients on their transplant

Table 5.4   Lymphocytotoxicity crossmatch methods

Standard crossmatch	Magnetic microsphere crossmatch
90 mins	25 mins
Separate T and B lymphocytes using density centrifugation and nylon wool or sheep erythrocyte rosetting	Separate T cells with anti-CD8 monoclonal. Separate B cells with anticlass II monoclonal coated magnetic microspheres
30 mins	Incubate cells and serum 40 mins
90 mins	Incubate with complement 40 mins
Add: Eosin Formalin to fix the reaction	Add: Acridine orange Ethidium bromide India ink EDTA to fix reaction

waiting lists to ascertain whether they have developed anti-HLA antibodies and if so, whether those antibodies are specific for a particular HLA antigen such as HLA-B8. To screen the serum it is 'crossmatched' with cells from between 20 and 100 normal individuals, either selected randomly or on the basis of their known HLA types, in order that a complete range of HLA types is represented in the panel. If the result of these screening 'crossmatches' are all positive with panel members that possess a particular antigen such as HLA-B8, and no panel member with HLA-B8 yields a negative result, it is possible to define a specificity for those panel reactive antibodies. In many patients with anti-HLA antibodies it is not, however, possible to define such a clearcut specificity.

Some patients, especially those who have at least two risk factors for sensitisation (previous pregnancy, blood transfusion, or failed transplant), develop antibodies to most HLA antigens and react positively with most or all cells on the panel. Such patients are termed 'highly sensitised' and present particular problems in renal transplantation (see below).

Different centres use the information gained from screening in different ways, but each attempts to use the screening result to predict the best donor for a particular patient. For example, in some centres, donors will be excluded if they are known to possess HLA antigens to which a patient has previously made a specific antibody. Patients whose sera have been screened carefully and show no positive reactions may not require a crossmatch with the actual donor, the centre relying instead upon the near certain prediction of a negative crossmatch by the screening result.

The B lymphocyte crossmatch   There are a variety of antibodies which cause cytotoxicity of B lymphocytes without a positive T lymphocyte crossmatch (Table 5.5). There remains considerable doubt about the degree to which antibodies causing a positive B lymphocyte crossmatch may influence the outcome of transplantation. There have been well recorded cases of successful transplantation despite the presence of anti-HLA class II antibodies and 'weak' anti-HLA class I antibodies, which may cause a positive crossmatch limited to B lymphocytes because of their higher density of class I molecules compared with T lymphocytes.

The degree to which centres currently use the B lymphocyte crossmatch varies widely, though most would assess living related donors with both a T and B lymphocyte crossmatch test.

Lymphocytotoxic auto-antibodies   It has been known since the early 1970s that patients with systemic lupus erythematosus may have antibodies which are cytotoxic to their own lymphocytes when incubated *in vitro* with rabbit complement in a standard crossmatch test. These 'autoreactive

Table 5.5   Causes of a positive B lymphocyte crossmatch

1.	HLA class II antibodies – anti-HLA-DR or -DQ
2.	Weak (low titre, low affinity) HLAclass I antibodies
3.	Autoreactive lymphocytotoxic antibodies
4.	Non-auto, non-HLA antibodies

antibodies', which may be found in 10% of patients awaiting renal transplantation, have a number of distinguishing characteristics (Table 5.6). The two most important characteristics, from a clinical viewpoint, are that auto-antibodies cause positive crossmatch tests but do not imply that a graft will be rejected. In other words auto-antibodies cause false positive crossmatches and thus deny some patients a successful transplant.

It is possible to use a variety of techniques to identify the false positive results due to auto-antibodies. In some centres, autologous lymphocytes are collected in large numbers either by cytopheresis of the patient or by *in vitro* culture of the patient's lymphocytes using EBV transformation. When a sufficient quantity of lymphocytes have been collected the autoreactive serum is incubated with them, to absorb all the antibody onto the surface of the lymphocytes and leave the remaining serum free of the autoantibody while retaining any true anti-HLA antibody. A less labour intensive method is to incubate the autoreactive serum with dithiothreitol (DTT) for 30 minutes at 37°C immediately before crossmatching, chemically to reduce IgM antibodies and thus remove autoreactivity. In order for this to be reliable it must first be ascertained that the patient does not have IgM anti-HLA antibodies which would also be removed. This may be done using a panel of cells derived from patients with chronic lymphocytic leukaemia since the auto-antibody will not react with these cells.

Augmented crossmatch techniques

There are two commonly used methods of detecting antibodies specific to allogeneic lymphocytes which are more sensitive than the standard crossmatch techniques described above. The major problem faced by all very sensitive tests is that of increasing the number of false positive results. These augmented techniques thus have to be validated against the standard crossmatch test. At the present there does not appear to be a substitute for the standard test for recipients of first transplants, while the more sensitive tests may have a role in the complicated and sensitised recipients of regrafts.

Antiglobulin lymphocytotoxic crossmatch   The antiglobulin test involves incubation of anti-human globulin or anti-kappa chain antibodies with the serum and lymphocytes before addition of rabbit complement. In this way it has proved possible to detect low titres of antibody binding to lymphocytes. The test requires careful attention to detail and has not been applied

Table 5.6   Characteristics of lymphocytotoxic autoantibodies

- Predominantly IgM (though IgG also reported rarely)
- Specificity or specificities unknown
- May react with T and B lymphocytes, or B lymphocytes only
- Cause positive crossmatches with all potential donors
- Are often more reactive at cold (4°C) than warm (22° or 37°C) incubation temperatures
- Do react with the cell line K562
- Do not react with lymphocytes from most patients with chronic lymphocytic leukaemia
- Do not predict renal transplant rejection

universally, though there are retrospective studies which indicate that it may provide clinically relevant results, especially for recipients of regrafts.

Flow cytometry crossmatch Flow cytometry (FC) is a technique which detects immunoglobulin bound to the surface of lymphocytes by measuring the binding of fluoresceinated anti-immunoglobulin antibody. The test is thus independent of both complement binding and cell lysis used as the marker of antibody binding in the standard crossmatch test. Early reports suggested that the FC crossmatch was several thousand fold more sensitive than cytotoxicity techniques. The current conclusion is that the increased sensitivity is more modest, in the order of two or three doubling dilutions. Increased sensitivity and the ability to detect non-complement fixing antibodies have yet to prove to be clinically relevant though, as with the anti-globulin technique, there are a number of studies which suggest a useful role for FC crossmatching in sensitised recipients.

Alternative crossmatch techniques

The tests described above are all designed to detect antibodies directed at HLA antigens present on donor T lymphocytes (class I) or B lymphocytes (class II). While these antigens are the prime candidates for the rejection response, the vascular endothelium, representing the first point of contact between recipient blood and donor tissue, has specific antigens to which an antibody response may be mounted.

Endothelial monocyte crossmatching A number of laboratories have identified antibodies which are directed against antigens present either exclusively on vascular endothelial cells (usually derived from umbilical cord or from donor aorta removed at nephrectomy) or on both endothelial cells and monocytes, but not on lymphocytes. While experimental studies have suggested a role for these antibodies, particularly in the rare cases of rapid rejection of HLA-identical living related grafts, few laboratories have established the techniques routinely.

Epithelial cell crossmatching A recent study has suggested that some patients may posses antibodies which can be detected by flow cytometry using an epithelial cell line (A549). These antibodies have been linked to an epidemic loss of grafts in one paediatric transplant unit, but there is no confirmatory evidence as yet.

Cellular assays In living related transplants it is possible to perform a standard mixed lymphocyte response test to determine the degree of class II mismatching. This may help to decide between one haplotype matched donors, if there is a choice, or if serological or genetic analyses are unable to give a clear answer. Lymphocyte mediated cytotoxicity assays, in which direct killing of donor lymphocytes by recipient cells can be measured, have received interest but are not currently used in routine clinical practice. Measurement of precursor cytotoxic lympho-cyte (p CTL) frequency is a technique being developed to assess the frequency of cytotoxic T cells directed at a particular donor's mismatched histocompatibility antigens. If the technique can be simplified, automated and made more rapid it may have application in determining the best recipient for a particular donor kidney, especially in patients who have rejected previous grafts.

Highly sensitised and regraft recipients

Patients who have developed multiple anti-HLA antibodies and those who have proved their propensity to reject one or more renal transplants in the past present a major challenge despite current immunosuppressive therapies. Many studies have shown that patients who reject a first graft rapidly have a higher risk of rejecting the second graft, compared to those who lost a first graft either from technical reasons or from chronic rather than acute rejection.

Tissue matching

The effect of HLA matching on graft survival is greater in recipients of regrafts than it is in first grafts. Figure 5.4 shows data from the Collaborative Transplant Study in patients treated with cyclosporin and demonstrates that the spread of graft survival rates with increasing antigen mismatches is greater in the second transplants than the first. It is thus even more important to match for HLA antigens in regrafts.

Centres vary in their approach to avoiding donor HLA antigens that particular patients have either previously been exposed to or demonstrated an antibody response to. The most restrictive position is; firstly, to avoid

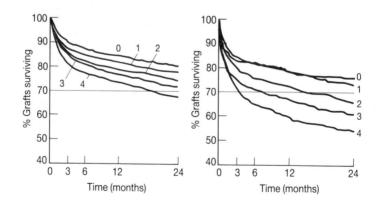

First cadaver grafts
0 MM n = 1685
1 MM n = 5723
2 MM n = 8087
3 MM n = 5348
4 MM n = 2222

Second cadaver grafts
0 MM n = 434
1 MM n = 1104
2 MM n = 1364
3 MM n = 886
4 MM n = 357

Figure 5.4   HLA-B, -DR mismatches in first and second cadaver transplants (data from the Collaborative Transplant Study 1990).

donors who have an HLA antigen which was present on the previous transplant (and mismatched with the recipient); secondly, to avoid donors with HLA antigens to which a previous antibody specificity has been shown on panel screening; and thirdly, exclude all donors with a positive crossmatch against serum taken at the peak of the patient's sensitisation, as discussed below. Some centres use all of these criteria, while less restrictive centres use none of them. Most, however, use one or two of these criteria in combination, from which it can be seen that there is no clearcut and correct choice.

Tissue matching of the first graft is known to have one effect that is agreed upon. Failure of a well matched transplant leads to lower levels of panel reactive antibodies (PRA) than failure of a poorly matched transplant. The other factor that probably affects the degree of sensitisation of patients awaiting a second graft is the use of blood transfusion during or after first graft failure. Nephrectomy of a failed graft and prolongation of immunosuppression have not been proven to reduce sensitisation though this is believed by some.

Peak positive, current negative crossmatch transplants

A positive crossmatch using serum taken at the time of a transplant (current) implies the presence of circulating antibodies reactive to donor antigens. The implication is that those antibodies will lead to an immediate, hyperacute graft rejection. The only known exception is when the crossmatch result is due to autolymphocytotoxic antibodies as discussed above. When the current serum crossmatch is negative but only (peak) serum samples taken months or years before are positive, the implication is rather different. In this situation the crossmatch provides information on the patient's previous immune responses and by implication their immunological memory, and may thus predict the response to the transplant. This assumption was shaken in the early 1980s, when it was shown, from Canada, that patients may receive a successful transplant if the current serum crossmatch was negative, even though the peak crossmatch was positive. There have since been many studies confirming that the peak crossmatch result is not relevant in sensitised recipients of first transplants. The situation is less certain when the recipient of a peak positive current negative crossmatch graft has become sensitised by loss of a first graft. Many centres would thus ignore a peak crossmatch result in first graft but not in regraft recipients.

Desensitisation

If it is possible to transplant patients whose PRA levels have fallen naturally with time, then it has been reasoned, it may also be possible to transplant patients whose antibody levels have been reduced by immunosuppression. In at least one series this has proved possible, with antibody levels in selected patients reduced by a combination of either plasmapheresis or adsorption of immunoglobulin onto a staphylococcal protein A column, combined with immunosuppression using corticosteroids, cyclophosphamide and azathioprine. Others have however not found it straightforward to repeat this success. Attempts to induce desensitisation by repeated blood transfusion or by use of immunosuppression without plasmapheresis have not yet been successful. The best treatment options for a highly sensitised patient thus still remain an extremely well matched graft or dialysis.

Sharing networks and tissue type matching

There are a multitude of systems for allocation of kidneys on the basis of HLA matching. The criteria for allocation vary from mandatory sharing only of zero HLA mismatched kidneys, to allocation on the basis of best HLA match.

The United Network for Organ Sharing (UNOS) in the United States has developed a mechanism for sharing kidneys across the country to maximise the number of completely matched transplants (by mandatory allocation of kidneys to fully matched recipients) and to minimise wastage of kidneys that cannot be used locally. A points system is used to redistribute kidneys, taking into account both the degree of HLA matching and the time that a patient has been awaiting transplantation.

In the United Kingdom the UK Transplant Service offers a similar mechanism for sharing of kidneys shown to have only a single HLA-A or -B antigen mismatch (i.e. 5/6 antigen match), and has the ability to direct kidneys not used by a local centre to the best tissue matched recipient in the UK. There are other examples of central registries and waiting lists where allocation is based upon HLA matching, such as Eurotransplant.

In Australia the allocation of kidneys is based upon a national matching service and computer system. The difference between the Australian system and most others is that crossmatching is performed in the centre where the donor organs are retrieved, for all patients in Australia. Allocation is based upon a points system with the greatest weighting on HLA-B and -DR antigen matching, a lesser effect from HLA-A, and influence given to time waiting and degree of sensitisation. One kidney goes to the recipient with the highest score in the country while the other goes to the recipient with the highest score in the state in which the kidneys have been retrieved.

Estimates have been made of the proportion of transplants that can be well matched with a given number of patients on a waiting list. In practice in Australia in a three year period in which 1563 first cadaver transplants were performed, 8.3% had no B or DR mismatches, 33% only one mismatch and only 3.8% a complete mismatch (Data from the ANZDATA Registry).

Consequences of tissue type allocation

The tissue type has the moral and ethical convenience of being a predetermined lottery which is not susceptible to the vagaries of physician bias and may thus be seen by some as the fairest mechanism for allocation of a scarce resource. The advantage of improved transplant outcome has been reviewed above, as has the reduced degree of sensitisation that accrues after failure of a well matched graft.

The issue that is now starting to receive considerable attention is the problem that reliance on tissue matching yields. In a hypothetical homogeneous population in which all tissue types were equally uncommon, then tissue matching would be the equivalent of giving each person on the waiting list a single lottery ticket and telling them to wait until their number comes up. In reality, some tissue types are very common, while others are rare and so some patients are given 400 lottery tickets while others may receive only one. The patients with the common tissue types are thus transplanted quickly. The situation is further complicated in multicultural societies such as Australia, or where there is a large ethnic minority such as in America, when organ donation is largely confined to the Caucasoid population. In this situation, because of the ethnic differences in tissue type, some people receive 400 lottery tickets, some one ticket and many of the ethnic minority no ticket at all. In other words accepting patients with tissue types that do not appear in the donor population is futile if tissue match is the only criterion and the pool of patients waiting is large. Allocation systems based upon tissue type matching have to become more sophisticated if the problems generated by them are to be overcome.

Further reading

1. Terasaki PI (ed). *Clinical Transplants*. Los Angeles: UCLA Tissue Typing Laboratory, 1990.
2. Gjorstrup P. Anti-HLA antibody removal in hyperimmunised ESRF patients to allow transplantation. *Transplant Proc* 1991; **23**, 392–395.
3. Kasiske BL, Neyland JF, Riggio RR *et al*. The effect of race on access and outcome in transplantation. *New Engl J Med* 1991; **324**, 302–307.
4. Gjertson DW, Terasaki PI, Takemoto S, Mickey R. National allocation of cadaveric kidneys by HLA matching. Projected effect on outcome and costs. *New Engl J Med* 1991; **324**, 1032–1036.
5. Bodmer JG, Marsh SGE, Albert ED *et al*. Nomenclature for factors of the HLA system 1991. *Tissue Antigens* 1992; **39**, 161–173.

6   The surgical procedure

Unlike the elective surgical procedure of living related transplantation, notification of a cadaver kidney suitable for a recipient on the transplant waiting list can come at any time of the day or night. Despite possible inconvenience to both staff and patient, organised activity along a predetermined pathway is required to provide a safe surgical transplant procedure for the recipient and optimal conditions for the donor kidney to achieve initial function.

Notification and acceptance

The designated clinician or transplant coordinator in the transplant unit will be the first person to be notified by either a regional organ sharing scheme or the local tissue typing laboratory. While the call may be expected because of recent local organ retrieval, in other instances the kidney may be offered by another transplant centre participating in an organ sharing scheme. Details of the condition of the donor and retrieval procedure may influence the clinician's decision to accept the kidney, or may influence selection of the initial immunosuppression regimen and recipient fluid management (Table 6.1). The information should be obtained routinely and compared with the requirements of the designated recipient.

Donor kidneys are a precious resource and there are very few absolute reasons to refuse an offer of a lymphocytotoxicity crossmatch negative cadaver donor kidney. More often than not, the donor kidney will be less than perfect. Warm ischaemia time of up to 30 minutes and total renal ischaemia time of cold stored kidneys of up to 48 hours are marginal but acceptable. There is no objective evidence that age disparity between donor and recipient should influence acceptance provided the donor is aged between three and 65 years. An experienced transplant surgeon can cope with most anatomical variations and shortcomings.

Organ sharing schemes usually have a time limit after notification for

Table 6.1 Essential donor kidney details

The donor
Name
Age
Donor institution
Tissue typing details
   – HLA -A, -B, -Cw, -DR, DQ (donor and recipient)
   – current and peak antibody crossmatch result
   – dates of crossmatch serum samples
Blood group
Cause of brain death
Days on ventilator
History of hypotension (BP < 80 systolic)
Use of inotropic drugs
Urine output in previous 24 hours
Urine output in hour prior to donor procedure
Viral status for HIV, CMV, hepatitis B and C
Past medical history
   – hypertension
   – diabetes
   – tumours
Destination of other kidney
Other organs retrieved

The kidney
Side
Number of arteries
Presence of aortic patch
Number of veins
Trauma to kidney
   – before retrieval
   – during retrieval
Ureteric abnormalities
Type of perfusion fluid
Quality of perfusion
Time of aortic cross clamping
Time of cessation of artificial ventilation
Warm ischaemia time
Cold stored or machine perfused

acceptance of the donor kidney of between 20 and 120 minutes, before offering the kidney to the next most suitable patient. Maintaining an accurate and frequently updated waiting list of healthy recipients is of utmost importance. The list should include current recipient contact telephone numbers, including those of relatives and neighbours, and other useful information, such as places of regular leisure or recreational activities.

The compliance of the potential recipient to the requirements of being on a transplant waiting list is often proportional to the quality of patient education provided by the transplant centre. They should be informed of symptoms and signs of medical conditions that may make them temporarily unsuitable for transplantation. For potential recipients with a particularly mobile lifestyle, access to a telephone paging system is worth consideration. Police and other community services are often sympathetic to the cause of transplantation and may be of help if the recipient cannot be found easily.

Once contacted and informed of the availability of a suitable cadaver kidney, the recipient is instructed to fast and requested to attend the transplant unit without delay. Arrangements for transportation of the donor kidney may be necessary. Clearly defined and reliable routes for shipment of the donor kidney should be predetermined and include the source and means of financial reimbursement for the various utilities involved.

Finally, the transplant ward nursing staff, surgeon and anaesthetist should be informed of the expected time of arrival of the recipient and the donor kidney, together with information regarding potential delays that may result if the dialysis dependent recipient has not been dialysed in the previous 36 to 48 hours.

Recipient preparation

When the patient arrives in the transplant unit, an initial clinical assessment of the patient is made, with careful attention to fluid status and the need for dialysis prior to transplantation. Because of the time constraints placed upon transplantation, the conventional approach to medical admission with sequential history, examination and investigation usually has to be abandoned in favour of an order based upon time factors (Table 6.2). Early notification to the anaesthetics department allows pre-operative assessment by the anaesthetists who often do not have the benefit of prior knowledge of the recipient.

Blood samples for measurement of serum electrolytes, full blood count and crossmatching of two units of blood together with chest X-ray and electrocardiogram are routine, and often have to precede history and examination. Although many of these investigations act as a reference point for changes that may occur after transplantation, early knowledge of electrolyte abnormalities is essential to assess the need for additional dialysis prior to transplant surgery. Correction of hyperkalaemia by temporary measures is unacceptable and dialysis prior to surgery, despite the delay involved, is preferable to emergency dialysis in the first 24 hours after transplantation.

The history of a patient on the transplant waiting list will usually have been well documented at their initial assessment for transplantation and should be known by both the transplant surgeon and physician. Emphasis should therefore be placed on the history of recent events, with evidence sought for current infections, recent onset of cardiac or respiratory disease and other new events that may preclude or delay transplantation. For example, if a blood transfusion has been given after the most recent serum sample used for crossmatching with donor lymphocytes, a repeat recipient lymphocytotoxicity crossmatch using current serum is advisable, particularly if the recipient is a parous female.

Oral immunosuppression given prior to the transplant procedure provides patients with a modification of the normal immune response to an allo-antigen. In most centres, the selected drugs are given early in the pre-operative work-up. Peritoneal dialysis fluid should be drained from the abdominal cavity, examined and sent for culture, following which the catheter may be capped or locked with dilute heparin. Diabetic patients are

Table 6.2 Pre-operative assessment of the renal transplant recipient on arrival in the transplant ward.

Peripheral blood investigations
Full blood count
Clotting profile (prothrombin ratio, activated prothrombin)
Electrolytes, creatinine, urea, blood sugar.
Liver function tests
Calcium, phosphate
Viral titres (CMV, herpes viruses, EBV)
(crossmatch 2 units of blood)

History
Establish history of transplant work-up
Recent infections
   – respiratory tract, ear, dental, skin, urine, peritoneal
Systems enquiry for current symptoms
Blood transfusions
Medications
Dialysis
   – timing of last haemodialysis
Social history including work arrangements after transplantation
Last oral intake of food, fluid, medication

Examination
Weight, temperature
Cardiovascular
Respiratory
Abdominal
Hydration status
CAPD exit site/arteriovenous fistula
Central nervous system

Other investigations
Chest X-ray
Electrocardiograph

best managed with a constant infusion of dextrose and an insulin infusion, the rate of which should be adjusted hourly according to capillary blood glucose meaurement.

The incidence of deep vein thrombosis in the first month after transplantation is only about 2%, presumably because of anaemia and the effects of chronic uraemia and haemostasis. For this reason, prophylactic subcutaneous heparin is usually not thought necessary prior to transplant surgery, nor during the time of cross clamping of the external iliac vessels during the surgical procedure. However, the use of recombinant erythropoietin and the higher incidence of vascular thrombosis in some units may alter attitudes towards more frequent use of prophylactic heparin.

Immediately prior to theatre, the recipient should be bathed or showered using an antiseptic soap. The site of vascular access should be clearly marked and demonstrated to the anaesthetic staff. Protection of the vascular access site is advisable using non-circumferential padding to reduce the risk of occlusion by inadvertent external compression by the careless surgeon or assistant during the surgical procedure.

Informed consent

The complications of renal transplantation are numerous but the risks are not unacceptable to the majority of potential recipients who are keen to obtain the improved quality of life provided by renal transplantation. Although careful preparation minimises these risks, the recipients' expectations of transplantation may be inappropriate, and it is therefore important to inform the potential transplant recipient on repeated occasions of the risks involved. Information can be provided by the referring nephrologist, the transplant surgeon and physician at the initial assessment for transplantation, by previously transplanted patients and by formal patient education sessions and information booklets. The recipient should therefore be well informed by the time he or she presents for hospital admission for transplant surgery. Frequently, however, the clerking resident medical officer witnesses the signature of the recipient on the consent form for transplant surgery. Despite the need for haste in preparing the recipient for the operating theatre, time should be set aside to outline again the risks and realistic expectations of transplantation, a suggested list of which is summarised in Table 6.3.

Anaesthesia

The advent of predictable and effective anaesthetic agents together with improved monitoring technology has made renal transplantation safer for a wider group of recipients.

Pharmacokinetics of renal failure

The altered pharmacokinetics of anaesthetic drugs in patients with chronic renal failure must be appreciated (Table 6.4). When there is a choice, drugs should be selected on the criteria that they are short acting and neither nephrotoxic nor renally excreted. For example, atracurium and vecuronium are non-depolarising and short acting muscle relaxants that are respectively spontaneously deactivated (Hofmann elimination) and 85% excreted unchanged in the bile. Potassium control may be crucial when the kidney does not function immediately and therefore solutions containing potassium should be avoided.

Intravenous access

Anaesthetists should be aware of the presence of any arteriovenous fistulae, particularly in relation to use of automatic blood pressure cuffs and peripheral intravenous cannulae. A triple lumen central venous line provides the best venous access, both at time of operation and in the early postoperative period, and facilitates simultaneous infusion of fluids and drugs, as well as monitoring of central venous pressures. The internal jugular vein is preferred to the subclavian vein for central venous access (Figure 8.1).

Table 6.3 Recommended issues for discussion with the renal transplant recipient before obtaining informed consent

1. *The kidney*
    The quality of the donor kidney, degree of tissue matching and basis upon which it has been allocated. Test results for virological risk markers such as hepatitis B and C, cytomegalovirus and HIV.

2. *The technical complication rate*
    The risks of technical problems after transplantation surgery may be up to 10% and include arterial and venous thrombosis, ureteric infarction and stenosis, urinary bladder leak and lymphocele.
    Graft loss as a result of these complications is less than 5%.

3. *Initial function*
    There is a 60 to 90% chance of initial transplant function, with a need for at least temporary dialysis if transplant function is delayed.

4. *Rejection episodes*
    There is a 50 to 80% chance of at least one rejection episode which will need treatment with extra medication.

5. *Immunosuppression medication*
    Stopping immunosuppression will result in rejection and loss of the renal transplant and so medication will have to be taken for life
    There are specific complications associated with the use of individual immuno-suppressive drugs. In particular they are:
    – hypertension and hair growth with cyclosporin
    – marrow suppression with azathioprine
    – skin fragility, diabetes, impaired wound healing with steroids
    – opportunistic viral infections with biological anti-lymphocyte preparations.
    There are also general complications associated with immunosuppression of which it is important to mention increased susceptibility to rejection and to certain types of cancer especially skin cancers (in fair skinned patients living in sunny climates)

6. *Graft survival*
    The one year graft survival for a cadaver transplant should be estimated based upon the transplant unit's own statistics and known risk factors for the individual such as whether it is a first or regraft, the degree of tissue type match and the patient sensi-tisation.

Table 6.4 Altered drug pharmacokinetics associated with chronic renal failure

1. Renal excretion dependent on glomerular filtration is negligible until transplant function is obtained

2. Hepatic metabolism is decreased, producing prolonged drug bioavailability

3. Anaemia produces a hyperkinetic circulation which increases absorption from subcu-taneous and intramuscular routes of drug administration

4. Decreased extracellular fluid space and protein binding reduces volume distribution and increases drug concentration in the central fluid compartment

Fluid status

Fluid administration during the transplant procedure is best guided by standard clinical parameters. It is particularly important to avoid inadequate cardiac filling pressures and low pulmonary arterial pressure, both of which have been demonstrated to affect adversely the chances for initial function of the transplanted kidney. Infusion of at least 1.5 litres of colloid during and soon after the transplant procedure has been demonstrated to provide better initial transplant function.

Blood transfusion

The cardiovascular system of patients with chronic renal failure is usually well accommodated to chronic anaemia, and unless the recipient is suscep- tible to angina or has a haemoglobin of less than 60 gm/l, transfusion prior to transplant surgery is not necessary and is probably best avoided. If blood transfusion is required during surgery, a white cell filter can be used, although for rapid volume replacement the filter is neither practical nor effective.

Analgesia

Renal transplantation is a painful procedure and adequate analgesia is necessary at the completion of the operation to reduce the risk of sudden patient movement that may inadvertently disturb the carefully positioned renal transplant. An intravenous opiate infusion makes for a smooth transi- tion between anaesthesia and the transplant ward. An epidural infusion of local anaesthetic agent is a practical alternative, provided hypotension resulting from lower limb vasodilation is avoided by adequate intravenous fluid replacement.

Preparation of the donor kidney

The cadaver kidney usually arrives packaged in at least two sterile contain- ers to facilitate transportation. It is the practice of most donor surgeons not to dissect the vascular supply of the kidney at time of organ retrieval. Hence, the kidney will still be surrounded both by perinephric fat and the adrenal gland superiorly (Figure 6.1). Preparative surgery must therefore be undertaken in a bowl of saline ice slush on a separate dissection table with the help of a surgical assistant.

Donor renal artery dissection

Firstly, the perinephric fat is removed from the lateral side of the kidney to assist with orientation and subsequent dissection of the vessels. Vascu- lar clamps are then placed at either end of the aortic patch and the renal artery identified. The renal artery is initially most safely approached from the posterior aspect, with care taken to avoid damage to accessory renal arteries and early renal artery branching. Branches to the adjacent lumbar muscles and adrenal gland require ligation. The longer right renal artery usually has at least one branch to the adrenal gland whereas the shorter left renal rarely has branches other than to the kidney (Figure 6.2a and b).

The extent to which dissection is taken into the hilum of the kidney is best determined by experience. Too little dissection will make the subsequent vascular anastomoses difficult, whereas too much dissection will compromise the blood supply to the renal pelvis and ureter, and after

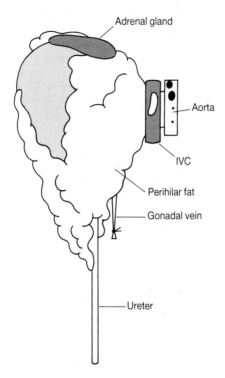

Figure 6.1 Anterior view of unprepared right sided cadaver donor kidney.

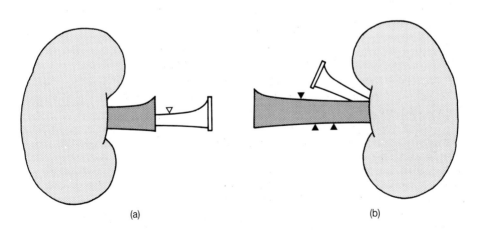

(a) (b)

Figure 6.2 Anterior view of completed vascular dissection demonstrating (a) right kidney with long renal artery and short renal vein; (b) left kidney with short renal artery and long renal vein.

revascularisation of the kidney, makes control of bleeding around the hilum of the kidney both arduous and hazardous.

Donor renal vein dissection

Vascular clamps are similarly placed at either end of the vena caval patch and the renal vein dissected towards the hilum. The right renal vein is short, usually between 2 and 4 cm in length. Multiple small branches are encountered passing into the hilum of the kidney and if care is not taken, these are easily torn despite minimal traction. If torn, the defects in the renal vein require repair with 6/O prolene sutures. Extra length of the right renal vein can be made more available by ligation and division of these small branches. The longer left renal vein usually has an adrenal branch placed proximally and superiorly, a lumbar branch posteriorly and a gonadal branch inferiorly (Figure 6.2b)

Donor ureter dissection

Completion of the removal of perinephric fat and the adrenal gland is undertaken after the dissection of the vascular supply. It is critical to the subsequent integrity of the ureter that its blood supply, which is dependent on branches of the renal arteries, is preserved. The chain of small anastomotic vessels that accompanies the ureter are more easily identified after revascularisation of the kidney. Debridement of the peri-ureteric tissue is therefore best completed after revascularisation of the renal transplant. Before transplantation, some debridement is possible by avoiding dissection in the triangular space between the tip of the lower pole of the kidney, the ureter and the hilum of the kidney. The pelvis of the kidney is rarely seen during this preparatory surgery (Figure 6.3).

Accessory renal arteries

Approximately 25% of cadaver donors have at least one kidney with more than one renal artery, half of which will have more than one vessel supplying both kidneys. As there is no collateral arterial circulation within the kidney, identification and preservation of the entire arterial supply is mandatory. Failure to do so may result in a necrotic segment of kidney and formation of a calyceal urine leak. Inferior pole accessory arteries or branches are of particular importance as they usually provide the arterial supply to the renal pelvis and ureter.

In most instances, the multiple arteries will branch from the aorta in close proximity, permitting use of an aortic patch of up to 3 cm in length for a single arterial anastomosis (Figure 6.4a). Alternatively, an accessory renal artery can be anastomosed to the side of the main renal artery (Figure 6.4b). Formation of a common orifice between two renal arteries is the least favoured option (Figure 6.4c), and may not be feasible with a right kidney because of the short arterial length. In such instances it may be necessary to perform an end-to-end anastomosis of the accessory artery to the inferior epigastric artery or a separate anastomosis to the recipient iliac artery.

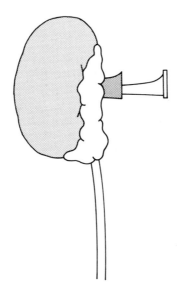

Figure 6.3 Anterior view of right cadaver kidney after removal of an appropriate amount of perinephric tissue and adrenal gland.

Accessory renal veins

Major accessory renal veins are infrequent although more common in right sided kidneys. Small tributory veins can be ligated without impairing venous drainage of the transplant. If, however, two or more renal veins of greater than 0.5 cm diameter are present, the venous anastomosis can be performed either using a patch of donor inferior vena cava or after creating a common lumen between the renal veins.

Ready for transplantation

Preparation of the donor kidney for transplantation is completed by fashioning aortic and vena caval patches. The renal artery is a thin walled vessel when compared to the iliac artery. Hence it is advantageous to use a circumferential patch (Carrel patch) of the donor aortic wall to reduce the potential risk of concentric stenosis at the site of anastomosis. A margin of aorta and vena cava of 1 to 2 mm around the perimeter of the orifice of the vessels is sufficient (Figure 6.5). Often, a short flap may be present in the orifice of the renal vein and require division. A small wedge biopsy may be taken from the renal cortex for later evaluation of possible acute tubular necrosis or for comparison with arterial changes seen in biopsies taken after transplantation.

 With long periods of warm ischaemia associated with non-heartbeating cadaver donors, many surgeons once advocated capsulotomy or division of the capsule from the upper pole to the lower pole around the lateral border. The reasoning behind this advice was the frequency of rupture of the cortex

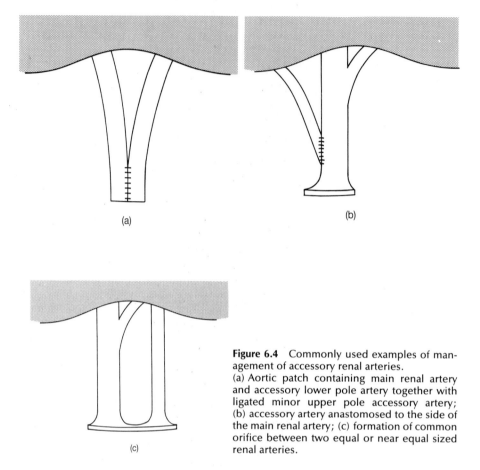

Figure 6.4 Commonly used examples of management of accessory renal arteries.
(a) Aortic patch containing main renal artery and accessory lower pole artery together with ligated minor upper pole accessory artery; (b) accessory artery anastomosed to the side of the main renal artery; (c) formation of common orifice between two equal or near equal sized renal arteries.

produced by swelling associated with acute tubular necrosis. Capsulotomy is not necessary with heartbeating cadaver donor kidneys. Many surgeons prefer at this point to package the kidney in an elasticated mesh to facilitate handling during formation of the vascular anastomoses (Figure 6.6). The mesh is then removed immediately before revascularisation of the renal transplant.

The operative procedure

Although not a technically difficult procedure, the operation of renal transplantation requires the services of an experienced surgeon to cope with the variable anatomy of the donor kidney and respond appropriately to unusual or unexpected events after revascularisation of the renal transplant. Indeed, it is these challenges, together with the sight of urine production within minutes of completion of the vascular anastomoses, that makes renal transplantation surgery both stimulating and rewarding.

The donor kidney is transplanted in a heterotopic position in one or other

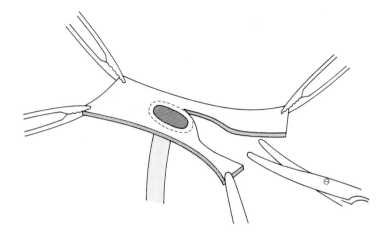

Figure 6.5 Fashioning of the aortic patch containing the orifice of the renal artery.

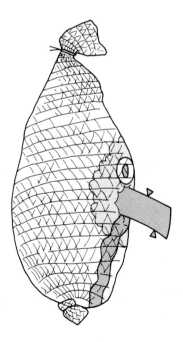

Figure 6.6 Prepared donor kidney in elasticated mesh ready for recipient venous anastomosis.

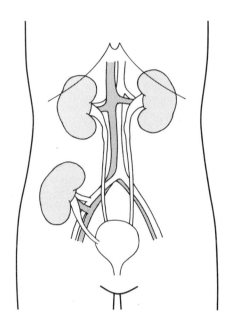

Figure 6.7 Heterotopic position of the renal transplant in the right iliac fossa. The transplant renal artery and vein are anastomosed to the recipient iliac vessels. The recipient's own kidneys remain *in situ*.

iliac fossa, with vascular anastomoses to the recipient iliac vessels and ureter anastomosis to the urinary bladder (Figure 6.7). It is more common and preferable for a single surgeon to perform the entire procedure, although in some institutions, separate specialist surgeons undertake the vascular and urinary tract anastomoses.

Planning the operation

If only by convention, it is customary to transplant the left donor kidney into the recipient right iliac fossa and the right kidney into the left iliac fossa. It is claimed that the ureter is more accessible when the donor kidney is transplanted into the contralateral side. Nevertheless, it is possible for either kidney to be transplanted into either side, with consideration being given to many factors including previous transplantation, previous urological surgery, the presence of a large polycystic recipient kidney, and the potentially difficult anastomosis of the right renal vein to the more deeply situated recipient left iliac vein. The left donor kidney is more easily transplanted because the relative lengths of the artery and vein are more appropriate to the anatomical position of the external iliac vein and artery (Figure 6.8a).

In contrast, the right kidney has a long renal artery which may increase the risk of subsequent kinking of the arterial blood supply to that kidney (Figure 6.8b). This risk can be minimised by proximal placement of the arterial anastomosis to the external iliac artery, or alternatively, short-

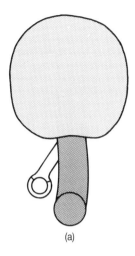
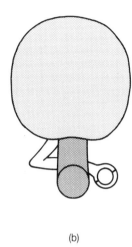

(a)   (b)

Figure 6.8 Cross-sectional views of configuration of a donor kidney artery and vein and recipient iliac vessels. (a) Favourable situation of a left sided kidney into the right iliac fossa; (b) less favourable situation of a right sided kidney into left iliac fossa with demonstration of potential for kinking of the donor artery.

ening the right renal artery and performing an end-to-end anastomosis to the internal iliac artery on that side. However, particularly in older patients, the patency of the internal iliac artery is unpredictable because of frequent atheromatous plaque formation. Furthermore, patients frequently require second and third renal transplants and in such instances, buttock claudication and impotence may follow if both internal iliac arteries have been used for end-to-end anastomoses.

Having planned which iliac fossa to use for the transplant procedure and administered a single intravenous dose of a broad spectrum antibiotic for infection prophylaxis, the recipient is catheterised. A 22 gauge silastic double lumened urinary catheter is inserted into the urinary bladder using an aseptic technique. The catheter is then connected to a 500 ml bag of normal saline containing a urinary antiseptic such as chlorhexidine. This fluid is used later to facilitate the ureteric anastomosis by distending the recipient urinary bladder. Difficulty may be encountered with catheterisation of older men because of urethral strictures or prostatic hypertrophy. Urethral dilatation of strictures can be attempted with care or alternatively, a safer option is to place a suprapubic catheter at the completion of the ureteric anastomosis. If the recipient fails a trial of voiding at the end of the first week after transplantation, urethral pathology can then be managed definitively.

Incision and mobilisation of iliac vessels

A curvilinear supra-inguinal incision is made from the midline, 2 cm above the pubic symphysis and passing laterally. Most of the incision is in a skin crease but the lateral extent of the wound will depend on the relative sizes of the patient and the donor kidney. The retroperitoneal plane is entered by splitting the external oblique muscle and aponeurosis in the line of its fibres, and division of the internal oblique transversus abdominus muscles in the line of the wound. Inadequate haemostasis when dividing these muscles is a frequent source of peritransplant haematoma. The anterior rectus sheath is divided in the line of the wound and the rectus abdominus muscle retracted medially, or divided if necessary. Further access to the retroperitoneal plane is afforded by division of the inferior epigastric vessels and mobilisation of the spermatic cord in males, or division of the round ligament in females. The external iliac vessels are displayed by sweeping the peritoneum and abdominal contents upwards and medially. Exposure is assisted by the use of a self-retaining retractor.

Adequate mobilisation and controlled exposure of the recipient iliac vessels is essential to successful vascular anastomoses. The shape and size of the human pelvis is immensely variable. Whereas some are favourable to retroperitoneal surgery, others, like that of the short muscular adult male, seem almost impossible even with a favourable left sided donor kidney. An appropriate length of external iliac artery, situated superficial and lateral to the external iliac vein, is mobilised, avoiding anteriorly situated lymphatic trunks, which should be divided only where necessary and only after ligation with non-absorbable suture material. If an end-to-end anastomosis to the internal iliac artery is required, full mobilisation of the distal common iliac artery, its bifurcation and the internal iliac artery to its point of trifurcation is required.

The external iliac vein is similarly mobilised. In a deep pelvis, particularly with a short donor renal vein, more complete mobilisation of the external iliac vein is possible by division of the internal iliac veins. However, division of these veins is not easy and haemostasis is difficult to obtain if an internal iliac vein is damaged or ligatures are inadequate.

Venous anastomosis

Whether using a right or left donor kidney, the venous anastomosis is most easily performed to the distal external iliac vein. Systemic heparinisation is not required before cross clamping the vein. Longitudinal venotomy of appropriate length is made and after careful orientation of the donor kidney, superior and inferior stay sutures of 5/O prolene are inserted followed by a triangulating suture to retract the lateral wall away from the medial wall. The anastomosis is then completed using a continuous over and over suture of 5/O prolene everting the edges (Figure 6.9).

Arterial anastomosis

Most surgeons prefer to anastomose the end of the renal artery and its aortic patch to the side of the external iliac artery. In most instances, the arteriotomy is made in the proximal portion of the external iliac artery or distal common iliac artery. Depending upon the diameter of the iliac artery, an elipse of anterior arterial wall may be removed.

A recommended suture technique for the end-to-side anastomosis, one which should first be perfected in an experimental model or an elective

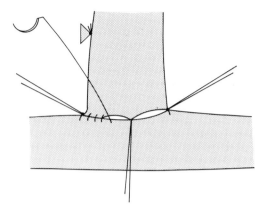

Figure 6.9 Anastomosis of the end of the donor kidney renal vein to the side of the recipient external iliac vein.

peripheral vascular surgery procedure, involves the use of a single continuous 6/O or 5/O prolene suture. About eight suture loops are positioned at the superior end of the anastomosis, somewhat like the strings of a parachute. The donor artery is then pulled down, thus providing accurate apposition to the recipient arteriotomy (Figure 6.10). The distal end of the patch is then accurately positioned by either trimming the patch or extending the arteriotomy as necessary.

If the aortic patch is not present, it is preferable to anastomose the end of the donor renal artery to the recipient internal iliac artery. This situation occurs with living related transplantation; if the aortic patch has not been retrieved; if the aortic patch is unsuitable because of atheroma; or when the right renal artery has been shortened because of the risk of subsequent kinking. The end-to-end anastomosis with the internal iliac artery is generally more time consuming, for it is advisable to insert interrupted sutures

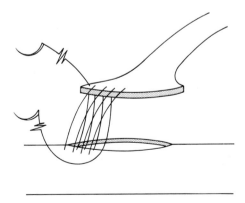

Figure 6.10 End-to-side anastomosis, using the 'parachute technique' of the aortic (Carrel) patch containing the donor renal artery to the recipient iliac artery.

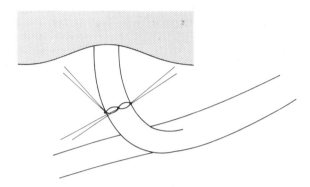

Figure 6.11   End-to-end anastomosis of the donor renal artery to the recipient internal iliac artery.

around at least one half of the diameter of the anastomosis to reduce the likelihood of subsequent suture line stenosis (Figure 6.11).

Excessive retraction on the renal vessels, particularly during a difficult or awkward anastomosis, may result in renal artery spasm or formation of an intimal flap and must therefore be avoided. Equally, appropriate haste is required to complete the vascular anastomoses in order to minimise warming of the kidney which increases exponentially after 30 minutes. For the same reason, handling of the kidney should be kept to a minimum and is facilitated by placing an arterial clamp on the packaging material in which the kidney has been placed. Regular cooling of the kidney with ice cold saline will reduce the warming effect produced by the operating theatre lights. The optimum combined vascular anastomosis time is between 25 and 35 minutes.

Removal of the vascular clamps is followed by rapid reperfusion of the renal transplant with a seemingly pulsatile increase in size. Small bleeding capsular blood vessels and hilar vessels, indicative of good transplant perfusion, often require diathermy and ligation respectively at this time. Urine production is usually seen within one to two minutes, and is assisted by maximum hydration (CUP 10–15 CmH$_2$O) of the recipient together with intravenous bolus doses of frusemide 1 mg/kg and mannitol 0.25 gm/kg.

Coping with the unexpected

Despite careful planning and attention to technical details, unexpected events occur that may be critical to the perfusion of the renal transplant. Calcified atheromatous plaques in the iliac vessels are common, particularly in the elderly recipient with a history of smoking. Endarterectomy of the recipient iliac artery is best avoided if possible, because of the risk of raising an intimal flap. If essential, it should be performed before removing the kidney from ice cold storage. Eversion endarterectomy of the internal iliac artery is less problematic.

Realisation that the renal transplant is upside down at the completion of the vascular anastomoses, although disappointing for the surgeon, is compatible with good transplant function provided the transplant ureter is of sufficient length to take a more circuitous route to the bladder. In this situation, use of a double J ureteric stent is advised to prevent kinking of the ureter. Revision of both vascular anastomoses would provide an unacceptable combined anastomotic time in excess of 60 minutes, but nevertheless may be necessary if the ureter is of inadequate length. If one or other transplant renal vessel is twisted or kinked, the single anastomosis can be revised or the vessel divided and resutured in an end-to-end fashion after removing a segment or re-aligning the vessel. Reperfusion of the renal transplant with recipient arterial blood before recognition of the unsatisfactory vascular anastomoses produces warm ischaemia, reduces the feasible time for revision of the anastomosis, and emphasises the need for careful orientation of the kidney before commencing the vascular anastomoses.

Despite technically sound vascular anastomoses, not all renal transplants promptly reperfuse with recipient arterial blood. Sluggish reperfusion can be the result of an inadequate cardiac output, poor donor kidney preservation or damage to the intima of the renal artery. A proximally situated intimal flap, if recognised early, can be managed by excision of the affected portion of the renal artery followed by an end-to-end anastomosis. Intimal flaps in the region of the hilum of the kidney are very uncommon, difficult to detect and repair and usually lead to the loss of the renal transplant. In the absence of an arterial injury, transplant perfusion may be improved by increasing the central venous pressure with colloid infusion, and commencing an intravenous infusion of dopamine at 2 to 4 μg/kg/min.

Good initial perfusion followed by deterioration of perfusion within minutes of removal of the vascular clamps can be indicative of hyperacute rejection (see Chapter 9). This antibody mediated event is difficult to reverse, and if the process is rapid, the renal transplant may be lost on the operating table, or soon after, as a result of intravascular thrombosis. The hyperacutely rejected kidney is often described as being blue and flabby in appearance and the diagnosis can be confirmed by cortical biopsy and frozen section.

The decision to remove a renal transplant on the operating table after revascularisation is often made with difficulty. At the time of operation, assessment of perfusion can be made by observation of bleeding following a small cortical incision or by temporarily occluding the renal vein and assessing the renal transplant for pulsatile increase in size. If after completion of the ureteric anastomosis, there is still evidence of arterial flow into the renal transplant, the transplant wound should be closed and blood flow assessed at the completion of the procedure by either an ultrasound duplex study or radionucleotide scan.

Ureteric anastomosis

The transplant procedure is completed by re-establishment of the urinary tract. The ureteric anastomosis should be secure, heal without risk of subsequent stenosis, and prevent reflux of urine back into the transplanted renal pelvis. The three available options are direct anastomosis of the ureter to the recipient bladder (ureteroneocystostomy), anastomosis of the ureter to the recipient ureter (uretero-ureterostomy), and anastomosis of the renal pelvis to the recipient ureter (ureteropyeloplasty). Although the latter two techniques may provide the optimal measure for prevention of reflux of

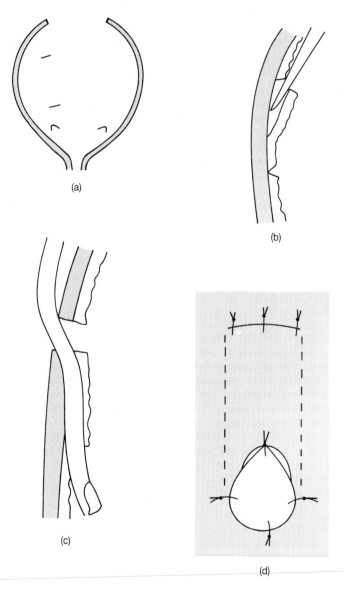

Figure 6.12 Intravesical uretero-ureterostomy. (a) Site of anterior cystostomy and the submucosal tunnel; (b) formation of the submucosal tunnel; (c) the transplant ureter is passed through a separate bladder incision and passed down the submucosal tunnel; (d) the completed ureter to bladder mucosa anastomosis.

urine into the transplanted kidney, they may necessitate recipient nephrectomy on that side and are associated with an unacceptable incidence of
anastomotic leak.

The recommended technique of implantation of the ureter into the
bladder was first described by Leadbetter and Politano in 1958 (Figure
6.12). Identification of the bladder is assisted by distention with saline
passed through the urethral catheter under pressure of gravity. An anterior
cystotomy of 4 to 5 cm in length is made to expose the lumen of the bladder.
Deaver retractors are placed in the bladder and a submucosal tunnel of
generous width and of 3 to 4 cm in length is directed towards a trigone of
the bladder (Figure 6.12a).

The optimum length of transplant ureter is the minimum sufficient to
achieve an anastomosis without tension, thus reducing the risk of kinking
and avascular necrosis of the end of the ureter. Active bleeding from the
end of the cut ureter is indicative of a healthy ureteric blood supply and
requires haemostasis by the placement of fine absorbable ties around the
bleeding vessels rather than diathermy. The end of the ureter is spatulated
for a distance of 1 cm, passed under the spermatic cord in males, through
a separate stab incision in the bladder wall and down the previously
fashioned submucosal tunnel, and anastomosed to the bladder mucosa
using interrupted 4/O absorbable sutures (Figure 6.12 b, c and d). The
bladder wall is closed in two layers with a continuous absorbable suture.
Use of a ureteric stent is seldom necessary.

The alternative extravesical ureteroneocystostomy, a technique described by Gregoir, is of particular value when the length of the ureter
available for anastomosis is short or the blood supply to the ureter is
dubious. The bladder is distended with saline and the muscle layers of
the anterolateral portion of the bladder divided to expose the mucosa for a
distance of 4 cm. The mucosa is opened at the distal end and the spatulated
transplant ureter sutured with interrupted absorbable sutures (Figure 6.13a).
The muscle layer is then repaired with interrupted absorbable sutures to
cover the terminal ureter (Figure 6.13b). This technique is less favoured by
some surgeons because of reported higher incidence of ureteric reflux, a

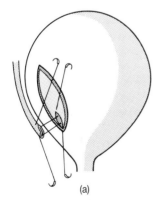

(a) (b)

Figure 6.13 Extravesical ureteroneocystostomy. (a) Anastomosis of the spatulated end of
ureter to the bladder mucosa; (b) the completed anastomosis with closure of the bladder
muscle layers over the transplant ureter to create a submucosal tunnel.

finding that is disputed by others.

Re-implantation of the ureter into the bladder can be associated with many technical difficulties. Duplex ureters occur in less than 1% and do not preclude transplantation. The common blood supply in the supporting tissue around the ureter is maintained, with the ureters passed into the bladder together, spatulated and either sutured together to form a common ureteric orifice or anastomosed separately to the mucosa. A small bladder or an inflamed friable bladder mucosa, however, is a more common finding, and makes construction of the submucosal tunnel difficult. Usually, the ureter can at least be positioned over a denuded area of bladder mucosa, and in time, mucosa will grow over the ureter and recreate a submucosal tunnel. In such instances, it is advisable to keep the urinary catheter in position for at least ten days after transplantation.

Positioning of renal transplant

Positioning the transplanted kidney is often difficult, particularly in patients who have a small deep pelvis or a shallow pelvic brim. The more usual

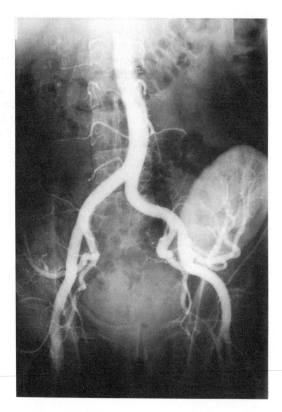

Figure 6.14 Angiogram of right sided donor kidney with two renal arteries on an aortic patch, anastomosed to the left external iliac artery. Note that the transplant has been positioned with the hilum directed laterally.

position is with the kidney placed on the psoas muscle with the hilum directed medially. However, the actual position of the kidney is less important than the need to prevent kinking of the transplanted vessels. The right kidney is more difficult to position especially if multiple renal arteries are present (Figure 6.14). Fixation of the renal transplant is of little value and may even precipitate kinking of the transplanted vessels or ureter.

A large closed vacuum drain is placed in the extraperitoneal plane adjacent to the anastomoses. Wound closure is in two muscle layers, with a continuous 1/O nylon suture, and followed by skin closure with a subcuticular nylon suture. In patients where poor wound healing is expected or those in whom an old wound has been re-opened, interrupted nylon sutures may be more appropriate. The surgical procedure usually requires two hours to perform and is followed by a routine bladder washout to remove any residual bladder blood clot.

Further reading

1. Calne RY. *A Colour Atlas of Renal Transplantation*. London: Wolfe Medical Publications, 1984.
2. Thrasher JB, Temple DR, Spees EK. Extravesical versus Leadbetter-Politano ureteroneocystostomy: A comparison of urological complications in 320 renal transplants. *J Urol* 1990; **144**, 1105–1109.
3. Ohl DA, Konnak JW, Campbell BA *et al*. Extravesical ureteroneocystostomy in renal transplantation. *J Urol* 1988; **139**, 499–502.
4. Moon MR, Roza AM, Johnson CP *et al*. Renal transplantation in patients with ileal conduits. *Clin Transplant* 1990; **4,** 370–375.
5. Churchill BM, Steckler RE, McKenna PH *et al*. Renal transplantation and the abnormal urinary tract. *Transplant Rev* 1993; **7**, 21–34.

7   Immunosuppression

Since the introduction of azathioprine in the early 1960s, the number of drugs which are capable of suppressing the immune system has increased steadily. These agents, either used alone or in combination, make it possible to prevent most patients from rejecting a renal transplant, without unacceptable side effects. Striking the balance between suppression of the immune response, drug specific side effects such as nephrotoxicity, and non-specific immunosuppression with the consequences of infection and malignancy is the challenge faced in each individual transplant recipient. While there is consensus on a number of general principles of drug usage, there are many equally viable strategies used by centres. The purpose of this chapter is to provide the information needed to prescribe immunosuppressive drugs in clinical practice and to demonstrate some of the ways in which they are used, in the certainty that each centre will have its own individual immunosuppression protocols.

The immune response and rejection

Control of the rejection response is at the heart of the success or failure of renal transplantation. An understanding of the immune response to allografted tissue is important to the management of clinical transplantation and immunosuppression.

Recognition that transplanted tissue is foreign, should probably be regarded as an unfortunate but relevant byproduct of the central role of immunity. The immune system functions to contain and eliminate invading micro-organisms. In order to achieve this, the immune effector cells have developed a mechanism of cellular interaction which involves both specific and non-specific molecular binding. Lymphocytes 'recognise' other cells using a specific molecular interaction between a receptor molecule (the T cell receptor – Ti on T lymphocytes, or surface immunoglobulin on B lymphocytes) and a molecule on the surface of the target cell, such as an epithelial cell infected with influenza. The molecule on the surface of the target cell has had to meet an exquisite design specification, namely the ability to bind with the T cell receptor and yet still convey the information that the cell is infected with one of an almost limitless number of possible alternative viruses. The design of the human leucocyte antigen (HLA) fulfills this specification, firstly by having a polymorphic (many alternative shapes) component that binds to the various alternative Ti molecules, and secondly by containing a groove on the uppermost surface, into which peptide fragments of the infecting organism can be bound (Figure 7.1). The T lymphocyte thus binds to its target cell through recognition of the virus peptide in the context of a standardised HLA binding molecule.

In the allograft transplant, the HLA molecules (together with any peptide fragments bound into their surface clefts) that are present on the cell membrane of the allograft cells are recognised by the host lymphocytes as 'infected' cells and thus destroyed. It is likely that the initial phase of the response involves an interaction between T lymphocytes bearing CD4 and CD28, and a specialised antigen presenting cell (APC) of either donor

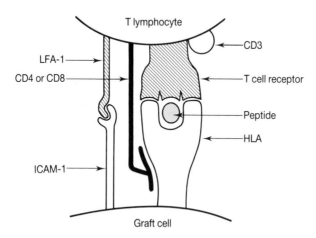

Figure 7.1 Interaction between a host T lymphocyte and graft cell. The interaction involves non-specific adhesion mechanisms such as that between LFA-1 and ICAM-1; molecule specific, CD4 for HLA class II and CD8 for HLA class I molecules; antigen peptide specific between the T cell receptor and the combination of HLA molecule and peptide: and a second signal such as that provided by CD28-B7 interaction. The CD3 molecule is responsible for triggering intracellular activation following binding of the T cell receptor and graft cell HLA molecules.

or host origin. Activation of CD4+ T lymphocytes (T_{helper}) and secretion of
IL2 subsequently lead to a cascade of cellular and humoral responses which
are directed at graft destruction (Figure 7.2).

Interleukin 2 (IL2) is a cytokine, or cellular messenger, which is central to
the process of T cell proliferation. IL2 is secreted by activated T cells and
stimulates proliferation only of T cells bearing the IL2 receptor. Normal
resting T cells do not express IL2 receptors. Within nine hours of being
exposed to antigen, however, T cells begin to transcribe mRNA for the IL2
receptor and thus become responsive to IL2 secreted into the local environment by activated T cells. It can thus be seen that IL2, secreted by and
acting on T helper cells which have recognised the allograft as foreign, is
the initiating and critical mechanism in graft rejection. The most effective
immunosuppressive drugs available in clinical practice act at this point.

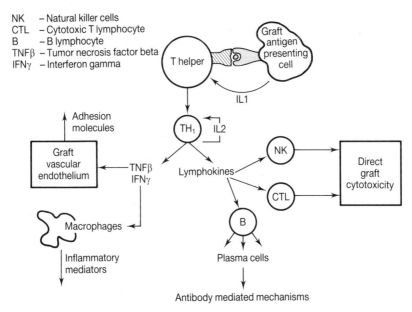

Figure 7.2 Diagrammatic representation of the cascade of recognition and effector mechanisms thought to be involved in the rejection response. The central process is activation and
proliferation of the T helper cell following recognition of the graft as foreign. This process is
dependent upon interleukin 2 secretion by the T cell, expression of the interleukin 2 receptor
on the T cell and autocrine stimulation.

Cyclosporin

Since its widespread introduction into clinical practice in 1983, cyclosporin (Sandimmun – Sandoz Basle – also known as cyclosporine, cyclosporine A and ciclosporine) has become the standard immunosuppressive
agent in transplantation. It is a natural peptide found in two strains of fungi
(Cylindrocarpon lucidum and Tolyprocladium inflatum) and noted to have
strong immunosuppressive potential *in vitro*. After evaluation in animal
models, it was first used in clinical practice in Cambridge in 1978 where

it was found to be effective but unexpectedly toxic. A number of clinical trials followed, demonstrating that cyclosporin improved renal graft survival by as much as 20% at one year. It was also responsible for making liver, lung, pancreas and cardiac transplantation practical and successful instead of experimental clinical treatments. Cyclosporin does, however, remain one of the more difficult drugs to use safely because of its narrow therapeutic margin, individual variation in absorption and metabolism, and multiplicity of drug interactions.

Mode of action

Cyclosporin inhibits early stages of lymphocyte proliferation in response to a variety of activation signals. The exact mechanism by which this is achieved has, however, yet to be clarified. It is known to have a strong inhibitory effect on synthesis by activated T lymphocytes of a variety of cytokines (IL2, IL3, IL4, gamma interferon and GM-CSF) at the level of mRNA transcription. This effect is probably caused by intracellular blocking of the cellular response to Ca^{++} influx, perhaps by interfering with activation of protein kinase C. How this is accomplished, except that it is likely to involve a specific cytoplasmic cyclosporin binding protein called cyclophilin, is now being elucidated.

Administration

Cyclosporin is available in three formulations, oral solution (100 mg/ml), capsules (100 mg, 50 mg and 25 mg) and intravenous solution (50 mg/ml). It is hydrophobic and requires maize oils (oral) or castor oils (iv) together with alcohol and emulsifying agents to disperse the drug in aqueous solutions. All formulations are best stored at room temperature; while the drug is stable below 30°C, the soft gelatine capsules start to melt above 25°C, but refrigeration is not recommended because of separation of the components. The oral solution has what is considered by most patients to be an unpleasant taste and if taken from styrofoam cups may be adsorbed onto the cup surface. It is best taken with a drink of milk both to disguise the taste and aid absorption. Many centres administer the oral solution or capsules once daily, though twice daily doses are used by others. There appears to be equivalence between the oral solution and capsules. In paediatric patients a large increase in dose when calculated by body weight, and moderate increase when calculated by surface area, is needed in comparison with adults. Almost all paediatricians would use twice or even thrice daily dosage in children. The intravenous solution has been reported rarely to induce significant anaphylaxis (probably due to the cremaphor carrier) and should be diluted in 20–100 ml of normal saline or 5% dextrose solution and infused over 2 to 4 hours twice daily, in a dose between 25 and 33% of the planned oral dose.

Pharmacokinetics

Measurement of cyclosporin levels in blood has some uses, since the therapeutic range is so small. Nevertheless, there is a tendency to believe that the resulting numbers mean more than they do, simply because they can be quantitated. Cyclosporin is absorbed with variable efficiency by different individuals but especially poorly by diabetics, with bioavailability varying between 1 and 70%. In the blood, only 1–2% is present as free drug with up to 50% bound to erythrocytes and leucocytes in a temperature dependent fashion, and the remainder bound predominantly to lipoproteins. Measurement of plasma cyclosporin levels is thus subject to the temperature of separation and for this reason, most centres use whole blood assays with a freeze-thaw cycle to lyse the cells. There are two types of assay available; those which measure only the parent drug, ignoring any metabolites (HPLC, and parent-molecule specific radioimmunoassay); and those which also embrace the metabolites (high performance liquid chromatography, non-specific radioimmunoassay and Abbott TDX). Cyclosporin is metabolised by the hepatic and gut cytochrome p450III systems to both immunologically active and inactive metabolites, which are then largely eliminated in bile. Renal failure and dialysis do not affect levels, but liver failure does.

Measurement of cyclosporin levels on a routine basis in all patients permits individual dose monitoring and leads to specific dose alterations on occasions. Table 7.1 details the clinical situations where the cyclosporin level contributes to a decision. The blood level, on its own, is seldom useful as a measure of the appropriate drug dose.

Side effects

The predominant concern with cyclosporin, apart from non-specific immunosuppressive activity, is nephrotoxicity. The range of side effects is, however, wide (Table 7.2). Hyperkalaemia, hyperuricaemia, hypertension and hyperlipidaemia cause concern to the physician, while hypertrichosis, gum hypertrophy and fine tremor understandably worry the patient most. The remaining side effects are generally rare. Nephrotoxicity, however, is to a greater or lesser extent almost universal when cyclosporin is used in maintenance doses above 1–2 mg/kg/day.

Cyclosporin is nephrotoxic in a variety of ways, some fully reversible and others not. It is a renal vasoconstrictor mediated by inhibition of vasodilatory

Table 7.1   The role of cyclosporin blood levels in clinical decisions *

Measurement of cyclosporin blood level is of particular value in the following situations:

1.   Drug interactions (see Appendix 2)

2.   Compliance and absorption when either are in doubt

3.   Declining renal function:
      Low level implies rejection more likely
      High level implies rejection less likely
      Neither can be relied upon without supporting clinical evidence

4.   Stable renal function and high level implies dose reduction is unlikely to precipitate rejection

5.   Abnormal liver function

---

*The number of variables involved and alternative tests available for cyclosporin measurement mean that the actual ranges have to be assessed in each laboratory.

prostaglandins, and to a lesser extent it is a tubular toxin. Four clinical manifestations of cyclosporin nephrotoxicity can be seen. Firstly, there is an increased incidence of initial poor or non-function and delay in recovery from acute tubular necrosis. For this reason many transplant centres delay introduction of cyclosporin until adequate renal function is observed. Secondly, there is a stable and reversible 'physiological' reduction of glomerular filtration rate (GFR) associated with vasoconstriction of glomerular afferent arterioles. The proposed mechanism for this action is impairment of vasodilator prostaglandin synthesis. This becomes apparent within ten days of the transplant operation and can be reversed if cyclosporin is stopped at any time in the first year at least. This improvement in GFR, observed when converting patients at three months, has been correlated with the blood level before stopping. Thirdly, there may be deterioration in renal function mimicking acute rejection and associated with morphological changes in the renal vasculature, glomeruli and tubules (see Chapter 10). Biopsy is often required to exclude rejection, and transplant function responds to dose reduction. Finally, chronic interstitial fibrosis

Table 7.2   Side effects of cyclosporin

Renal	**Decreased GFR***
	Hyperkalaemia
	Hyperuricaemia
	Hyperchloraemic metabolic acidosis
Cardiovascular	**Hypertension**
	Hyperlipidaemia
Hepatic	Transient enzyme
	Hyperlipidaemia
	Biliary stones
Pancreatic	Pancreatitis
	Diabetes mellitus
Haematological	Thrombocytopaenia
	Anaemia
	Thrombo-embolism
	Haemolytic uraemic syndrome
	Auto-immune haemolysis
Gastro-intestinal	**Gum hypertrophy**
	Anorexia
	Nausea and vomiting
Neurological	Grand mal seizures
	Fine tremor
Dermatological	**Hypertrichosis**
	Skin thickening
	Rash/flushing
	Brittle fingernails
Miscellaneous	Fluid retention
	Contraceptive failure
	Fetal nephrotoxicity
	Raynaud's phenomenon
	Anaphylaxis (iv preparation)

*The commonest side effects are printed in bold type.

with tubular atrophy, sometimes seen to be in stripes across the kidney, is the hallmark of irreversible chronic nephrotoxicity. Whether or not this is progressive or represents chronic rejection is subject to debate (see Chapter 11).

There have been a number of optimistic approaches to reduction of nephrotoxicity, of which dose reduction is probably the only effective strategy. Since prostaglandin infusion is effective in the rat, there were hopes that misoprostil or enisoprost (oral prostaglandin E_1 analogues) or Hydergine would show significant effects. Sandoz (Basle) have also researched the possibility that a less nephrotoxic analogue of cyclosporin A could be produced, but cyclosporin G, which has undergone clinical trials for this reason, has not been widely adopted so far.

Azathioprine

6-Mercaptopurine was the most successful of the early experimental immunosuppressive agents used in the late 1950s. Azathioprine was the result of a search for a less myelotoxic derivative and was introduced into clinical practice in 1962. In combination with corticosteroids, it remained the pre-eminent prophylactic immunosuppressant until the mid 1980s when it was supplanted by cyclosporin. Its importance has since re-emerged in combination with both cyclosporin and corticosteroids and it is probably still prescribed to the majority of transplant patients.

Azathioprine is inactive and requires metabolism in the liver to 6-mercaptopurine and other thiopurines, which are then excreted in the urine. It has non-specific effects upon the immune system. It is known to inhibit both DNA and RNA synthesis and can also block interleukin 2 production in the mixed lymphocyte response. Both oral and intravenous administration need only be once daily. The dose does not need to be altered in renal failure, and is not affected by dialysis. In regimens that do not use cyclosporin, it is usual to aim for azathioprine doses of 2.5 mg/kg/day, while with triple therapy including both cyclosporin and azathioprine, 1.5 mg/kg/day is the target dose (Table 7.3). Attempts have been made to measure 6-mercaptopurine levels by HPLC and bioassay. While there are suggestions that individual variations in drug handling could be important, in practice, the azathioprine dose is reduced only when enforced by leucopaenia or other specific complication. The single major drug interaction is with allopurinol, which by inhibiting xanthine oxidase interferes with breakdown of azathioprine and thus leads to accumulation of active metabolites. In the presence of allopurinol, the azathioprine dose has to be reduced by at least 50%. The common practice if both drugs must be used is to start azathioprine at 25 mg/day, increasing slowly towards 50% of the protocol dose unless prevented by onset of leucopaenia (Table 7.3). It should be noted, however, that when a dose of 1.5 mg/kg is used, the usual cause of leucopaenia is infection with cytomegalovirus.

The major side effect of azathioprine is its myelotoxicity, with leucocytes affected more commonly than platelets. Megaloblastic change in erythrocytes is frequent though it does not usually lead to anaemia. If anaemia is present without megaloblasts, the possibility of co-existing iron deficiency needs to be considered. Use of azathioprine does not of course prevent either co-existing B_{12} or folate deficiency. Hair loss is a concern to about 20 % of patients on higher doses and other side effects are uncommon (Table 7.4).

Corticosteroids

Neither azathioprine nor corticosteroids alone are capable of providing satisfactory immunosuppression, but when used in combination for the first time in the early 1960s they made renal transplantation a practical treatment. The three corticosteroid preparations commonly used in transplantation are oral prednisolone, intravenous hydrocortisone and intravenous methylprednisolone (for dose equivalence see Table 7.3). The effects of corticosteroids that make them useful in renal transplantation are complex and multiple. Their powerful anti-inflammatory action profoundly alters the effector phases of graft rejection, including macrophage function, seen especially when high doses are used to reverse acute graft rejection. Steroids have more specific effects in the afferent limb of the immune

Table 7.3 Azathioprine and corticosteroid dosing

1. *Azathioprine*
'High dose protocol' used in conjunction with low dose corticosteroids
    2.5–3.0 mg/kg/day for the first two years
    2 mg/kg/day for long term maintenance

'Low dose protocol' used as a component of triple or quadruple therapy
    1.5 mg/kg/day

Allopurinol–azathioprine combination
    25 mg/day maximum starting dose
    increase in 25 mg steps to 50% of protocol dose every two weeks

Azathioprine dose monitoring

Total WCC (x 10⁸/l)	Neutrophil (x 10⁸/l)	Platelet count (x 10⁹/l)	
> 4.0	> 1.0	> 100	No dose change
3.0–4.0	0.5–1.0	50–100	Decrease by 50 mg, recheck daily
< 3.0	> 0.5	< 50	Stop, recheck daily, reintroduce when levels rise

2. *Corticosteroid dose*
Relative potency compared with cortisol

	Glucocorticoid action	Mineralocorticoid action
Hydrocortisone	1	1
Prednisolone†	4	0.25
Methylprednisolone	5	0*
Dexamethasone	25	0
Fludrocortisone	10	300

† Prednisolone equivalent to prednisone
* 0 = mineralocorticoid action minimal

Table 7.4   Side effects of azathioprine and corticosteroids

Azathioprine
Bone marrow suppression
Megaloblastic change
Hepatic dysfunction (chronic veno-occlusive changes)
Acute pancreatitis
Hair loss
Skin thinning and fragility
Malignant tumours

Corticosteroids

General
Cushingoid facies and body fat distribution
Obesity (appetite stimulation)
Growth retardation in children

Dermatological
Skin fragility and thinning
Spontaneous bruising
Acne
Poor wound healing

Gastrointestinal
Peptic ulceration

Pancreatitis
Acute pancreatitis
Diabetes mellitus (peripheral insulin resistance)

Cardiovascular
Hypertension

Neurological
Posterior lentricular cataracts
Psychiatric disturbances–depression, psychosis

Orthopaedic
Avascular necrosis
Osteoporosis

---

response, inhibiting interleukin 1, preventing T lymphocyte proliferation and altering lymphocyte response to antigen at an intracellular level.

Initial regimens used in renal transplantation concentrated on the use of high doses of corticosteroids, with intravenous methylprednisolone immediately before the operation followed by 100 mg/day for the first two weeks, tapering to 20 mg by three months. It was shown, however, that similar results could be obtained by lower doses, commencing at 30 mg/day and reducing to 20 mg by three months. Large scale prospective randomised comparison of low and high doses has not been performed, and though small studies have shown significant reduction in side effects without compromising patient or graft outcomes, both regimens continue to be used in combination with either azathioprine or cyclosporin. Widespread use of low dose steroids, as a component of triple or quadruple therapies, has taken the heat from this debate.

Renal function does not alter dosage, but intravenous methylprednisolone boluses should be given after and not before haemodialysis in a dialysis dependent patient. Since bolus injection has been reported to cause cardiac arrhythmias and asystole it is advisable to administer high dose intravenous methylprednisolone over a minimum of five minutes, and it is thus most safely given by automated delivery over 30 minutes. Abnormal liver function significantly prolongs the half-life of prednisolone which is inactivated in the liver. By contrast phenytoin, phenobarbitone and rifampicin (by hepatic enzyme induction) significantly shorten the half-life. High oral doses are usually divided twice daily, while single daily doses are given in the morning to mimic the normal diurnal pattern of cortisol release.

Suppression of the normal adrenal axis implies that dose reduction must be gradual to avoid an Addisonian response, manifest as hypotension and general malaise. This is more of an issue after failure of a graft, months or years after a transplantation, than it is in dose adjustment with a functioning graft. The rate of change in dose, when the total daily oral dose is greater than 20 mg/day, is guided solely by the risk of rejection. Between 20 and 10 mg/day, dose reduction as fast as 2.5 mg per week can usually be tolerated without Addisonian side effects, while steps of 1 mg per fortnight are more reasonable when the dose is below 10 mg. Slow reduction is generally wiser than rapid changes to avoid allograft rejection; thus many units would use 1 mg steps for dose reduction.

Maintenance levels of corticosteroids between 5 mg and 15 mg/day, or an equivalent alternate day schedule, are used in the long term. Trials of immunosuppressive protocols involving cyclosporin have shown that steroid free regimens are possible.

Corticosteroid use in children presents problems, particularly because of its inhibitory effect on bone growth in children who are already of short stature as a result of their uraemia. Achieving normal renal function while the child is still prepubertal provides the best opportunity for obtaining a reasonable adult height and so the emphasis is on steroid reduction, alternate day therapy and steroid free protocols. Counteracting slow growth velocity with use of recombinant human growth hormone must still be regarded as experimental in view of early reports of unexpected graft rejection in children treated with growth hormone (see Chapter 15).

The side effects of corticosteroids, some of which are detailed in Table 7.4, can become the major preoccupation of both the patients and their physicians. The higher dose protocols, especially when large quantities of high dose intravenous steroid have been used, are associated with long term osteoporosis and avascular necrosis. Cataracts may require up to 10% of patients to have lens extractions and diabetes mellitus may become apparent for the first time or become more difficult to control in 10 to 20%. The appetite stimulant effect of steroids, together with lifting of previous dietary restriction, and a return to normal well-being between four and eight weeks after a transplant, may lead to steady weight increase until obesity intervenes. It has, however, been possible to reduce some of the side effects, such as peptic ulceration, by use of lower dose steroids and more liberal intervention with H_2 receptor antagonists, especially in patients with a previous history of peptic disease.

Antilymphocyte preparations

The potential for using the humoral response of a horse to injected human lymphocytes as a mechanism for control of allograft rejection, was recognised in the early 1960s. Advances in the definition of lymphocyte populations and then in production of selected monoclonal antibodies with specific molecular targets during the late 1970s and the 1980s have led to increasing popularity of what are known as the 'biological' agents.

Polyclonal antilymphocyte preparations

Human lymphocytes or purified thymocytes can be injected into a horse, goat, rabbit or other suitable animal to induce a polyclonal cytotoxic antibody response. Serum from these sensitised animals can then be purified and the resultant antibody preparation absorbed using red cells and platelets to remove unwanted and inadvertent toxicity from antibodies crossreacting with those cells. Antilymphocyte serum (ALS) or globulin (ALG) or antithymocyte globulin (ATG) is then available (after suitable safety testing) for intravenous administration. The major disadvantage of these preparations is the lack of standardisation between products from different manufacturers or even from batch to batch within a single product. A number of transplant centres manufacture their own (e.g. Minnesota and Sydney Universities) and there is a choice of commercial preparations (e.g. ATGAM, Upjohn; ATG, Fresenius and Pasteur-Merrieux).

These preparations are presumed to work by eliminating circulating T lymphocytes responsible for graft destruction. Lymphopaenia occurs following the first few doses, but the leucocyte count may recover without loss of immunosuppressive activity.

Administration is exclusively by intravenous injection. Prior to injection, testing of an individual's response may be recommended by the manufacturer, either by *in vitro* red cell crossmatching (especially with locally made ALG) or by subcutaneous test doses of 0.1 ml, observing for a hypersensitivity reaction of wheal and flare. Premedication with intravenous hydrocortisone and antihistamines is used before the first dose by some centres. The dose of ALG or ATG is dependent upon the individual preparation with a fixed dose being the standard approach (Table 7.5). Attempts to monitor the effect of ALG or ATG and adjust doses accordingly have not been successful.

The duration of treatment has varied from a few days up to as long as six weeks. Long periods of treatment are associated with more profound immunosuppression and risk of lymphoproliferative disease, particularly with repeated courses or when OKT3 (vide infra) has also been used. The usual maximum treatment length is in the order of two weeks. Side effects of polyclonal preparations can be divided into those seen immediately after injection and those seen in the medium to long term (Table 7.6). All of the former and some of the latter are directly attributable to the use of a foreign immunoglobulin, while the remainder are a consequence of immunosuppression rather than the particular form of polyclonal lymphocyte preparation.

Monoclonal antilymphocyte preparations

Only one monoclonal antibody preparation is available commercially for the

Table 7.5   Recommended doses of antilymphocyte preparations

Preparation	Prophylactic dose	Therapeutic dose for acute rejection
ATG–Fresenius	2–5 mg/kg/day 9–14 days	3–5 mg/kg/day 14 days
ATGAM–Upjohn	15 mg/kg/day 14 days	15 mg/kg/day 14 days
OKT3–Jannsen	2.5–5 mg/day 10–14 days	5 mg/day 10–14 days

Important Notes
* Testing for presensitisation by intradermal injection is advised by most manufacturers
* Premedication with intravenous corticosteroids and antihistamines is advised for at least the first dose
* Injection should always be given in a hospital with access to emergency resuscitation equipment in the event of anaphylactic reaction
* Dose reduction of other immunosuppressive drugs is advised by some manufacturers
* Recommended doses may vary from country to country. **Always check the local manufacturer's dose recommendations and package insert before prescribing**.

treatment of acute rejection at the present time (anti-CD3, orthoclone OKT3, Muromonab-CD3 Jannsen-Cilag Pharmaceuticals). There are, however, numerous monoclonal antibodies being investigated in individual centres, which are either directly cytotoxic to specific T lymphocytes (e.g. Campath-1) or act as target cell selectors and carry a conjugated toxin such as ricin or diphtheria toxin which causes cell lysis.

Monoclonal antibodies differ from the polyclonal preparations in that they contain antibody produced by a single carefully selected cell line. In essence, a mouse has been immunised with human T lymphocytes, spleen

Table 7.6   Side effects of antilymphocyte preparations

	ALG/ATG		OKT3
Immediate			
	%		%
First dose chills and fever	80	First dose pyrexia chills	90
Erythema pruritis	20	Dyspnoea and wheeze	30
Local thrombophlebitis	30	Chest pain	14
Anaphylaxis	1	Tachycardia	10
Haemolysis	1	Hypertension	8
Diarrhoea, vomiting	15	Hypotension	6
Serum sickness	10	Tremor	13
		Diarrhoea, vomiting	19
		Neurological*	1.7
		Fatal pulmonary oedema**	rare
Medium to long term			
Herpes simplex	50	Herpes simplex	30
Thrombocytopaenia	40		
CMV and other opportunistic infection	20	CMV and other opportunistic infection	20
Neoplasia	?	Neoplasia	?
Antibodies to ALG/ATG	10–50	Antibodies to mouse Ig	30

NOTE: *   Aseptic meningitis, seizures, confusion
  **   Fatal pulmonary oedema was seen in the first series of patients treated and was attributed to a combination of fluid overload and OKT3. Fluid states must be carefully assessed to prevent this complication.

cells from that mouse have then been fused with a myeloma cell line to ensure their perpetual growth, and the resultant clones tested for antibody production. The clones that produce antibody to the cell or molecule under consideration are then either cultured *in vitro* or re-injected into the peritoneal cavity of mice where antibody rich ascites is produced.

OKT3 is a monoclonal antibody directed at the CD3 antigen on the surface of T lymphocytes, the function of which is to act as a transducer of signal from the T cell receptor to intracellular activation events (Figure 7.1). CD3+ T lymphocytes disappear rapidly from the circulation following a dose of OKT3. T lymphocytes that subsequently return to the peripheral blood while OKT3 is present do not express either the CD3 antigen or the associated T cell receptor. These are functionally blind and thus ineffective for promulgating rejection of the transplanted kidney. Despite use of OKT3, CD3+ T lymphocytes can be found in plentiful numbers in renal transplant biopsies taken from OKT3 treated patients. It is thus not entirely clear whether destruction of T lymphocytes and modulation of the CD3 antigen fully explains the mechanism of action of OKT3.

Administration of OKT3 leaves patient and physician in no doubt that it has been given, since within a few hours of the first dose up to 90% of patients get a syndrome of fever, chills, rigors and general malaise. This response is thought to be due to release of tumour necrosis factor from lysed T lymphocytes and is best managed symptomatically and by premedication with intravenous antihistamines and high dose corticosteroids. It is critical to determine the fluid status of the patient before giving OKT3, since fatal pulmonary oedema occurred in three of the initial group of 107 patients treated. In practice this needs clinical examination, a chest X-ray and knowledge that an adult patient's weight is not more than 3 kg above their normal dry body weight. In view of the potential for severe reaction the first dose must be given by a doctor who is fully aware of the potential reactions and equipped to resuscitate the patient. The first dose should always be given in a hospital with intensive care facilities. By contrast, subsequent doses are often entirely uneventful and may be given on an outpatient basis.

The major side effects of OKT3 are detailed in Table 7.6, the least predictable of which is the 2 to 5% incidence of aseptic meningitis. Some patients may complain of lesser manifestations with only a mild headache, whereas others develop confusion, neck rigidity, photophobia and fever, suggestive of meningo-encephalitis. The neutrophil cell count and protein content of the cerebrospinal fluid (CSF) may be elevated, but culture is negative. Since it is impossible to exclude infection without CSF culture, this is mandatory. The symptoms and signs usually regress without intervention, whether or not the OKT3 has been discontinued.

Experimental immunosuppressive agents

An increasing number of agents known to be active immunosuppressants are at various stages of laboratory and clinical evaluation. In renal transplantation, the problem that has to be faced is that of proving that a new agent is superior, unless it offers a dramatic improvement in the profile of side effects. Current immunosuppressive protocols yield one year first cadaver graft survival rates in the region of 80 to 90%. If a new agent were to improve on these results by reducing loss of grafts from rejection, then a prospective randomised trial of up to 800 patients would be needed to have a reasonable chance of proving a 5% increase in graft survival at one year. Elimination of the remaining graft losses would require reduction in losses from technical problems at the time of surgery and death from myocardial disease and infective causes. New agents therefore face a difficult and expensive route to clinical acceptability.

A number of the potential immunosuppressive agents have been listed in Table 7.7, some of which have received only laboratory attention while others are proceeding through pilot clinical appraisal.

Lymphoid irradiation

Irradiation was one of the first methods of immunosuppression attempted in renal transplantation. There was limited success with the strategy of whole body irradiation since toxicity rendered it unacceptable. By contrast, total irradiation of the lymphoid system (TLI) has been extensively studied in experimental models of transplantation and has been shown to induce long lasting tolerance of the allograft. Clinical application, in at least three centres, has shown that recipients acquire tolerance of the graft, judged by suppression of the mixed lymphocyte reaction. Additional immunosuppression has included initial antithymocyte globulin and corticosteroids, but in some patients, all medication has been withdrawn successfully. The disadvantages are that the risk of viral infection is unchanged; reported graft survival rates of 75% are lower than achieved with cyclosporin; and the requirement for nine weeks of radiotherapy followed by early transplantation. TLI has thus not been widely adopted, but clinical research continues. The final mode in which radiation therapy is used in transplantation is use of local radiotherapy to the kidney undergoing allograft rejection. The logistics of this procedure have limited widespread use, as has the effectiveness of simpler steroid treatments.

Table 7.7 Experimental immunosuppressive agents with potential use in renal transplantation

1.	Total lymphoid irradiation
2.	FK506
3.	Rapamycin
4.	RS-61443
5.	15-Deoxyspergualin
6.	Brednin
7.	Photopheresis
8.	Immunoconjugated toxins
9.	Brequinar sodium

FK506

Use of FK506 was, until recently, limited to the laboratory and a single clinical centre, but it is now commercially available in Japan. It is a macrolide antibiotic extracted from *Streptomyces tsukubaensis* with potent immunosuppressive properties. The mode of action appears to be very similar, if not identical, to that of cyclosporin, binding to a cytosolic phosphatidyl propyl isomerase, and interrupting intracellular signal transduction. *In vitro*, FK506 selectively inhibits secretion of cytokines (IL2, IL3, IL4, gamma interferon) and IL2 receptor expression. Short courses in animal transplant models are capable of prolonging most solid organ allografts and may also control rodent models of auto-immune disease. In clinical practice it has been shown to be more effective than cyclosporin in liver transplantation but convincing data for renal transplantation have yet to be presented.

The major disadvantage of FK506 is its toxicity with severe vasculitis in dogs and some primates, though not the rat. In man the profile of toxicity has yet to be fully reported, though it is clear from the initial data that nephrotoxicity, diabetagenicity and neurotoxicity pose problems. FK506 has significant hepatotropic effects, an advantage in liver transplantation, but interacts with cyclosporin, leading to high levels of cyclosporin if the two are used in combination. One *in vitro* study has also suggested that the two drugs may be antagonistic if used in combination, though others disagree. Clinical application of FK506 now awaits the results of prospective randomised studies that have been established in North America and Europe. It would be reasonable to say that there is more expectant interest from liver than renal transplant units.

Other immunosuppressive agents

Rapamycin is a macrolide antibiotic with similar structure to FK506. Its function *in vitro* has been shown to be distinct to that of both cyclosporin and FK506, but it may bind to the same cytosolic protein as FK506. Clinical experience is limited at present.

A further drug in the process of early clinical testing is RS-61443, a pro-drug of mycophenolic acid, which blocks proliferation of both T and B lymphocytes and prevents generation of cytotoxic T lymphocytes. In animal models it prolongs graft survival without overt toxicity, while in the clinical situation RS-61443 appears to be promising in combination with cyclosporin.

Two alternative strategies may, in the future, prove to be successful in the clinic. Photophoresis involves psoralen sensitised ultraviolet irradiation of peripheral blood leucocytes leading to 'programmed cell death'. Early experimental and clinical data suggest that it may be effective in reducing graft-versus-host disease in bone marrow transplantation and in preventing heart transplant rejection. The second strategy, of considerable interest in oncology, is to conjugate toxins to a monoclonal antibody as discussed above. This 'magic bullet' has, except for treatment of hairy cell leukaemia, yet to leave the laboratory and further progress is eagerly awaited.

Prophylactic regimens

The aim of prophylactic or baseline immunosuppression is to provide a level of anti-rejection therapy that is tolerated over the long term but able to prevent overwhelming allograft rejection in most patients. It is expected that supplementation with high dose treatments for short periods of time will be needed, in a proportion of patients, to reverse episodes of acute rejection.

Initiating regimens

A number of alternative prophylactic regimens are detailed in Table 7.8, with the range of starting doses usually used. Azathioprine and prednisolone, the standard or conventional therapy from 1962 to 1983, continues to be the standard only for patients unable to afford cyclosporin and for a limited number of HLA-identical living related graft recipients. The balance between high steroid/low azathioprine doses and low steroid/high azathioprine doses has been discussed previously. The conclusion is in favour of the low steroid treatment, but ensuring that azathioprine doses are at least 2.0 or 2.5 mg/kg/day, to maintain good long term therapy.

The treatment options that use cyclosporin vary, depending firstly upon the number of drugs used in combination and secondly upon when cyclosporin is started and stopped. Because of the propensity for cyclosporin to both precipitate and prolong acute tubular necrosis (ATN), many centres avoid cyclosporin until a diuresis has been established, preferring to commence with azathioprine and prednisolone either with or without an anti-lymphocyte preparation for the first ten to 14 days. Therapy during the first two weeks may thus comprise one of the following regimens:

- cyclosporin
- cyclosporin, azathioprine and prednisolone
- ALG/ATG/OKT3 with or without azathioprine, prednisolone and cyclosporin, azathioprine and prednisolone.

After the initial period, most centres use some form of triple therapy with cyclosporin, azathioprine and prednisolone at moderate doses such as suggested in Table 7.8. At these doses of azathioprine and prednisolone, drug related side effects are relatively few, with bone marrow intolerance quite unusual unless precipitated by CMV or use of an additional agent.

Sequential or quadruple regimens which use prophylactic antilymphocyte preparations, have been suggested as the best option for sensitised and regraft patients. Evidence to support this claim is limited and grafts are lost despite heavy initial immunosuppression. The evidence as to whether OKT3 is more or less effective than ATG/ALG is contradictory. In general, however, it is thought reasonable to use these regimens for higher risk transplant recipients, if for no other reason than to reduce the incidence of early acute rejection and thereby perhaps reduce the overall corticosteroid dose.

Maintenance treatments

Azathioprine will usually be continued unchanged for long term mainte-

Table 7.8   Clinical Regimens for Immunosuppressive Drugs

Prophylactic regimens:

Starting doses	Cy mg/kg	Aza mg/kg	Pred mg	ATG mg/kg (days)	OKT3 mg (days)	Comment
Regimen:						
No cyclosporin		2.5	100			High dose steroid
		2.5	30			Low dose steroid
High dose Cy	17.5					
Double therapy	10–15		20–30			May convert to Aza/Pred after 3 months
Triple therapy	8–12.5	1.5	20–30			Delay start of Cy till graft function
Quadruple/ Sequential	8–12.5	1.5	20–30	1.5–2.0 or (7–14d)	5 (5–10d)	Introduce Cy after 7–14 days of ATG in sequential treatment

Cyclosporin Dose Reduction Schedules:

	Initial (mg/kg/day)	Dose reduction	Maintenance (mg/kg/day)
Single therapy:	17.5–15	Scheduled reduction to 10 by day 90	5
Double therapy:	10–15	5–8 by day 90 adjust by blood levels	5
Triple therapy:	8–12.5	5 by day 90	3–5

Alternative Regimens for Treatment of Acute Rejection:

	Oral high dose prednisolone	Intravenous methylprednisolone	OKT3	ALG/ATG	Plasmapheresis
Severe rejection	200 mg daily for 3 days followed by taper over 5–7 days	1 gm daily for 3–5 days	5.0 mg/day	Doses as per manufacturers instructions (see Table 7.5)	Usually reserved for vascular rejection. 3–5 day exchanges of 2–3L
Moderate rejection	120 mg daily for 3–5 days followed by taper over 5–7 days	500 mg daily for 3 days	2.5–5.0 mg daily for 5–14 days		
Mild rejection	120 mg daily for 3 days returning to baseline dose	250–500 mg daily for 3 days	Not applicable		

* Cy = Cyclosporin, Aza = Azathioprine, Pred = Prednisolone, ATG = Antithymocyte Globulin nance. Prednisolone dose reduction after the first three months may either be to a maintenance value between 5 and 15 mg/ day or equivalent alternate day dosing, especially in children, or may be stopped altogether. Early withdrawal of prednisolone, within the first two weeks after transplantation, has proved to be a consistent failure, but late withdrawal on the basis of currently available evidence is a viable option in selected patients. Steroid withdrawal may most logically be targeted at children and diabetic patients. Cyclosporin withdrawal has been undertaken by numerous centres with conversion from cyclosporin alone, cyclosporin and prednisolone, or triple therapy, to azathioprine and prednisolone maintenance. The incentive to withdrawal is the long term effect of cyclosporin on renal function and in particular the concern over progressive interstitial fibrosis of the kidney. The problem with conversion regimens is the appearance of late and unpredictable rejection in up to one third of those stopping cyclosporin. The alternative of low dose maintenance cyclosporin at perhaps 2 mg/kg has proved more attractive, if the continuing costs can be borne.

Treatment of acute rejection

Following diagnosis of acute allograft rejection, the ideal therapy would provide immediate and, complete reversal of rejection without subsequent episodes. All the treatments available fail in one or more of these aspects. A frequent response to a clinical diagnosis of acute rejection, or even 'probable acute rejection', has been to treat the patient first and ask the questions second. While this approach to declining renal function in patients treated with azathioprine and prednisolone alone may be reasonable, in all other regimens, it is more appropriate to ask the questions first by undertaking investigations, including biopsy, to provide a more secure diagnosis before instituting treatment.

Corticosteroids

Bolus corticosteroids in high doses, either using an oral protocol or intravenous methylprednisolone, form the mainstay of treatment of acute rejection (Table 7.8). There are firm adherents to both approaches. There has been a tendency to reduce the daily dose of intravenous methylprednisolone from 1 gm towards 250 mg. This, together with the increasing number of alternative treatments available to rescue rejecting allografts, has probably contributed to a reduction in avascular osteonecrosis and subsequent need for joint replacement surgery.

Antilymphocyte preparations

Polyclonal and monoclonal antilymphocyte preparations have both proved effective in reversing acute rejection, either as a 'rescue' treatment when steroids appear to be failing, or when used as first line treatment.

OKT3 therapy is complicated by the need to ensure the patient is not overhydrated and by the likelihood of reactions to the first dose. This means that treatment should not be instigated unless adequate and experienced medical and nursing staff are on duty. Up to 25% of patients treated for acute rejection with OKT3 will have a rebound rejection episode five to ten days after the end of treatment. Fortunately, this second rejection episode usually responds to intravenous methylprednisolone, although occasionally ATG may be used. Three myths about the use of OKT3 need exposing. Firstly, OKT3 is capable of reversing acute vascular rejection as well as cellular rejection. Secondly, the dose of maintenance immunosuppression does not have to be reduced, though it may be useful to stop cyclosporin for a few days, under the cover of OKT3, to reassess the contribution of cyclosporin nephrotoxicity. Thirdly, OKT3 can be used more than once, but induction of anti-mouse antibody limits its effectiveness on the second treatment in about 30% of patients and this second course probably leads to an unacceptable risk of developing lymphoma. Commercially available kits may be used to measure anti-OKT3 and anti-mouse antibodies in individual patients to predict the effectiveness of a second course. OKT3 has also been used at reduced doses (2.5 mg instead of 5 mg) and for shorter durations (five days instead of ten) in many patients.

Randomised prospective comparisons of OKT3 with the variety of antilymphocyte preparations are limited. There has been little to choose between them in terms of graft outcome, but side effects have been different. In one study, the infective risk was similar, but OKT3 led to severe first dose effects and pulmonary oedema; while in another, viral infection was significantly greater with ATG than OKT3. OKT3 does have the advantage of uniformity, while batch to batch variation of ALG/ATG may perhaps explain some of these contradictory trials.

Plasmapheresis

The role of plasmapheresis in treatment of acute rejection has been limited to use when graft histology demonstrates a predominantly vascular rejection. The implication is that humoral immunity is contributing significantly to the process and thus even temporary abrogation of the antibody response is useful. Controlled trials are sparse and unimpressive, so the decision to use plasmapheresis currently rests with the clinician who may be influenced by personal experience or anecdote.

Plasmapheresis has also been used as a prophylactic agent in patients planned to receive ABO incompatible grafts and when there has been an attempt to desensitise highly sensitised patients (Chapter 5).

Individual problems with immunosuppression

Pregnancy

Return of fertility to a previously uraemic and infertile patient carries the responsibility for provision of adequate contraceptive advice for both male and female. In females, oral contraception is prone to failure because of the interaction with cyclosporin (Appendix 2). The co-existing risks of thrombosis and hypertension also make this a less desirable option than intra-uterine devices or barrier methods. While pregnancy is certainly inadvisable in the first year after transplantation, views differ on later pregnancy. There have been many live births to both male and female transplant recipients using azathioprine and prednisolone immunosuppression. The risk of teratogenicity is confined to an increased risk of cleft palate. Cyclosporin, however, affects fetal renal function, and although reports have suggested full reversibility, one must be concerned about long term effects on the child's kidneys. It is also worth noting that cyclosporin is secreted in breast milk at a level sufficient to advise against breast feeding.

Paediatric patients

The major issues with immunosuppression in children are compliance, dose adjustment, and growth potential. These are discussed in detail in Chapter 15. Compliance with cyclosporin oral solution, because of its taste, can usually be resolved by addition to chocolate flavoured milk. The capsules tend to be rather large for the smaller children and one centre has recently introduced cyclosporin into the chocolate centre of a wafer biscuit. Dose adjustments have been discussed with respect to each drug individually. In Chapter 15 a series of alternative protocols for use in small children are outlined, designed to accommodate the higher metabolic rate and rapid inactivation of drugs.

Growth potential in a child with endstage renal failure is limited by uraemia and a functioning graft is thus a prerequisite for growth. This fact must be taken into account when considering steroid dosage. It would be the general aim of most centres to stop prednisolone entirely in paediatric transplant recipients. However, loss of graft function may cost more centimetres of height than the prednisolone would ever have cost. Nevertheless, it is clear that the greater the dose of corticosteroid, especially if given on a daily instead of alternate day basis, the less is the chance of growth. Individual decisions have thus to be made, in which the degree of graft match and early problems with rejection are balanced against the initial height and growth potential of the child.

Further reading

1. Thomson AW (ed). *Cyclosporin. Mode of Action and Clinical Applications*. Dordrecht: Kluwer Academic Publishers, 1989.
2. Carpenter CB. Immunosuppression in organ transplantation. *New Engl J Med* 1990; **323**, 1224–1226.
3. Thompson AW. FK-506: profile of an important new immunosuppressant. *Transplant Rev* 1990; **4**, 1–13.
4. Strober S, Dhillon M, Schubert M *et al*. Acquired immune tolerance to cadaveric renal allografts. A study of three patients treated with total lymphoid irradiation. *New Engl J Med* 1989; **321**, 28–33.
5. Hanto DW, Jendrisak MD, McCullogh CS *et al*. A prospective randomized comparison of prophylactic ALG and OKT3 in cadaver kidney allograft recipients. *Transplant Proc* 1991; **23**, 1050–1051.
6. Clark MGB, Rigden SPA, Haycock GB, Chantler C. Renal transplantation in children In: *Reviews 1*. PJ Morris, NL Tilney (eds). New York: Grune & Stratton, 1987.

8 Early management of the uncomplicated transplant recipient

Perhaps a quarter of the transplant recipients will have an uncomplicated renal transplant. Ideally, as recipients they would not be presensitised to histocompatibility locus antigens (HLA) and apart from endstage renal disease, would be free of other medical problems such as hypertension, diabetes and coronary artery disease. By choice, the donor kidney would be from a two haplotype matched sibling or otherwise, a left sided, HLA compatible, cadaver kidney retrieved from a haemodynamically stable, young, heartbeating donor at the recipient's own hospital. The surgical procedure would be undertaken within 24 hours of the cadaver donor procedure and without difficulty, preferably on a Monday morning. Within minutes of completing both the transplant vascular anastomoses in less than 35 minutes, transplant urine would start flowing. Maintenance immunosuppression would subsequently prevent significant acute rejection and would not be associated with either specific or non-specific side effects.

On the morning of the seventh day after transplantation, the uncomplicated transplant recipient would be pacing up and down the transplant ward, bags packed, cyclosporin capsules under his or her arm and requesting politely to be discharged from hospital, with arrangements made for daily outpatient clinic attendance. On the day of discharge from hospital, the serum creatinine would be in the range of 80 to 140 μm/l and the glomerular filtration rate (GFR) more than 60 ml/min. In contrast to a complicated transplant recipient, he or she would probably be forgotten within a few weeks by the staff of the transplant ward.

So where are the challenges of renal transplantation? A quarter of the recipients will have no primary renal transplant function, will require at least temporary dialysis and expect a one year graft survival of 10 to 20% less than that achieved by patients that do have primary renal function. At least 60% of the transplant recipients will have at least one episode of acute rejection requiring treatment, and about 5% will have a technical complication requiring further surgical intervention. Whereas these challenges are the subject of subsequent chapters, this chapter outlines the objectives of early management which are to create optimal conditions for primary transplant function, adopt prophylactic measures to minimise complications and provide careful surveillance to identify renal transplant dysfunction. The uncomplicated transplant recipient makes his or her way home on day seven or soon thereafter, not by good fortune but by good management incorporating clearly defined protocols, teamwork and attention to detail.

From operating suite to transplant ward

The transplant recipient is extubated in the operating suite and if in a stable respiratory and cardiovascular state, can be transferred directly to the transplant ward where appropriate equipment and facilities should be available for high dependency nursing (Table 8.1). Essential equipment includes that needed for cardiorespiratory resuscitation, an oxygen source and reliable suction. For the recipient, the advantage of direct transfer is placement in the care of experienced nursing and medical staff who are not only capable of monitoring the cardiorespiratory status, but also the progress of the recently transplanted kidney. The anaesthetist may remain with the transplant patient for a short period of time in the transplant ward and provide the transplant unit staff with a verbal as well as written summary of the relevant events during the surgical procedure. Operative intravenous fluid requirements, blood loss, urine production, ongoing analgesia and complications during the procedure are all important information that may influence the early care of the transplant recipient.

The structural ward facilities provided for the care of the transplant recipient are much debated and vary widely, depending upon individual

Table 8.1 Optimal facilities and equipment for high dependency nursing of renal transplant recipients

1. Positive pressure air conditioned room with single accommodation
2. Cardiorespiratory resuscitation trolley
3. Oxygen source
4. Reliable suction
5. Cardiac monitoring
6. Oxygen pulsoximeter
7. Intravenous fluid infusion pumps
8. Wash basin for medical and nursing staff
9. Private shower and toilet
10. Easy access to haemodialysis facility

unit preferences and whether or not they were built specifically for the purposes of transplantation. In the early 1960s, when transplant recipients received total body irradiation prior to transplantation, it may have been appropriate to provide stringent reverse barrier nursing facilities. However, with more specific immunosuppression now in use, transplant recipients are no longer profoundly immunosuppressed at the time of transplantation. Prevention of cross infection remains important and is best provided by single rooms with provision for individual shower and toilet facilities. Staff directly in contact with the patients must wear clean or disposable gowns and wash their hands before and after examining or attending to the needs of the recipient. It could be argued that even these measures are unnecessary and that hand washing alone is appropriate, as it is for examination of all hospital patients. However, nurses and medical staff do forget and these recommended measures are at least a reminder to attending staff that appropriate care should always be taken to minimise cross infection.

There are two possible scenarios on return to the transplant ward: either chaos or calm. Chaos could include a patient writhing round the bed in pain and complaining vehemently of the need to pass urine; the surgeon anxiously down on one knee milking every drop of urine out of the catheter with one hand whilst holding the patient down with the other; the physician searching for the jugular venous pulse whilst ducking and weaving around projectile vomit and sputum; the medical resident attempting to take blood from an arm whilst the nurse is either taking a blood pressure or searching for the mid point of the axilla; and the radiographer positioning the patient and asking for the room to be cleared whilst a chest X-ray is taken. In the meantime, whilst all this is happening at once, bemused relatives will be peering through the door of the room looking for signs of life from the recipient.

The situation of organised calm is clearly preferable, with the transplant recipient returning to the care of two experienced nursing staff and an experienced member of the medical team, whilst the surgeon informs the waiting relatives of the progress thus far. Intravenous lines can be sorted out and attached to infusion pumps, the surgical drains and urinary catheter placed in a position where they can be readily viewed, adequate analgesia provided and baseline measurement of parameters of cardiorespiratory and renal transplant function commenced. Baseline bloods are taken for serum biochemistry and haemoglobin, and a portable chest x-ray to assess the position of a central venous line.

Routine monitoring

The first 24 hours are critical for the renal transplant recipient with early management designed to provide an optimal physiological environment for renal transplant function, and to recognise quickly and manage appropriately any complicating factors. Intensive monitoring of parameters of cardiorespiratory function and those specific to the transplanted kidney are required in this period (Table 8.2). Thereafter, the frequency of monitoring should be appropriate to the medical condition of the recipient and the function of the renal transplant.

There is great temptation for the transplant team to follow the course of the transplant procedure by merely observing the numbers produced by routine monitoring and to believe that function of the renal transplant implies that there are no other problems. Equally important is a clinical assessment of the patient's state of hydration by evaluation of skin turgor, peripheral perfusion, JVP, peripheral oedema and chest auscultation. Auscultation of the renal transplant often detects a soft bruit in the early hours after transplantation and regular palpation can detect changes in tenderness and size of the renal transplant. The safety of the early period of aggressive hydration of the renal transplant patient is also dependent on continuous observation, allowing appropriate management decisions to be made on a given set of numbers and clinical findings. For example, hourly measurement urine does not obviate the need for continuous observation of urine for sudden reduction in flow or onset of haematuria, either of which may require urgent attention.

Table 8.2   Routine monitoring for the first 24 hours after renal transplantation

---

1. *Cardiopulmonary function*
   Hourly measurement of blood pressure, pulse and respiratory rate
2. *Hydration*
   Central venous pressure
3. *Renal transplant function*
   Hourly urine output or more frequently for polyuria
4. *Drain tubes*
   Nature and volume of fluid
5. *Other*
   Hourly temperature
   Six hourly blood sugar level by finger prick
   Patency of vascular access
6. *Laboratory specimens*
   Haemoglobin, serum electrolytes, urine electrolytes at six hourly intervals

---

Fluid management

Early renal transplant function has many advantages and may not be achieved unless the transplant recipient is adequately hydrated and normothermic. Assessment of the adequacy of hydration requires skill and experience, but a central venous access with a triple lumen catheter is essential (Figure 8.1). The internal jugular vein is the preferred site because of lower risk of pneumothorax at the time of insertion and the possible need for use of the subclavian veins for haemodialysis access.

Aggressive hydration

A central venous pressure (CVP) of 10 cm of H_2O without the assistance of artificial ventilation can be used as a guideline, although there is considerable patient to patient variation, depending on cardiac function, blood pressure and compliance of the peripheral vasculature. Patients treated by peritoneal dialysis and those haemodialysed in the 24 hours prior to transplantation are more likely to be fluid depleted. If urine production is poor, and depending on clinical assessment, it is reasonable to increase the CVP to 15 cm H_2O by use of bolus infusions of a colloid intravenous fluid such as plasma protein fraction, crystalloid or even blood transfusion. Also, an infusion of dopamine (2 to 4 $\mu g/kg/min$) should be started in an attempt to improve renal transplant perfusion by reducing renal small vessel vasoconstriction. Diuretics are usually of little or no value. After 48 hours, it is common for an adult patient to weigh 3 to 6 kg more than at time of transplantation.

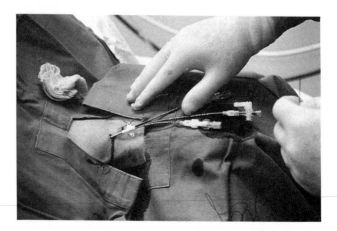

Figure 8.1 Placement of a triple lumen central venous catheter into the left internal jugular vein performed by the anaesthetist before commencement of the renal transplant procedure.

Maintenance fluids

Maintenance intravenous crystalloid fluids can be provided by a combination of 0.9% saline and 5% dextrose in a suggested ratio of two parts saline to one part dextrose, depending on urinary sodium loss and dextrose load. The hourly volume of infused fluid is best linked to the urine output, CVP and estimated insensible losses. A recommended formula is shown in Table 8.3. In the first hours after transplantation, urine output in excess of 1 litre/hour is often encountered and usually decreases spontaneously without the need to reduce the rate of intravenous fluid infusion. If the urine output remains in excess of 600 ml/hour, it may be appropriate to reduce the infusion rate with care to avoid rapid dehydration. In situations of high urine output, it is also necessary to check the serum and urine biochemistry at six hourly intervals to guide replacement of sodium, potassium, bicarbonate and calcium. Large volume infusion of 5% dextrose may lead to hyperglycaemia.

Blood transfusion

Most patients will present for renal transplantation with a haemoglobin between 60 and 100 gm/l, depending on the cause of their underlying renal disease and the use of synthetic erythropoietin. Furthermore, it is not uncommon for the haemoglobin concentration to fall by 10 to 30 gm/l as a result of haemodilution produced by aggressive hydration of the recipient. Physiological adjustment to the state of chronic anaemia implies that there is usually no need to transfuse the patient during transplantation surgery, unless excessive blood loss occurs. The same applies after transplantation, especially since a functioning renal transplant will slowly correct anaemia over the first one to three months after transplantation. Factors which favour transfusion include ongoing blood loss and symptomatic coronary artery disease. There is a theoretical advantage in using a white cell filter if transfusion is necessary.

Oral fluids

A nasogastric tube is seldom needed and best avoided, even though a degree of small bowel ileus is not uncommon after extensive retroperitoneal

Table 8.3 Maintenance intravenous fluid replacement in the first 48 hours after transplantation in adult recipients

CVP (cm of H_2O)	Volume replacement per hour
> 15	Previous hours' urine output
10–15	Previous hours' urine output + 30 ml
5–10	Previous hours' urine output + 70 ml
< 5	250 ml colloid bolus and previous hours' urine output + 70 ml

surgery, particularly if complicated by haematoma formation. Bowel sounds are usually audible on the first day after transplantation and oral fluids can be introduced and increased as tolerated so that intravenous replacement may be stopped by day three or four after transplantation. However, if urine production is greater than 200 ml/hour and the recipient is not overhydrated, continued intravenous supplementation is indicated to avoid dehydration that may be associated with impaired tubular concentrating ability of the renal transplant.

Postoperative analgesia

Renal transplantation is a painful surgical procedure, particularly in the lateral region of the wound where abdominal wall muscles have been divided. In addition, a large sized urinary catheter can be uncomfortable, and bladder spasm is common. Without adequate analgesia, the transplant recipient will become restless and sudden uncoordinated movements may dislodge intravenous lines, drains and even lead to kinking of the transplant vascular pedicle or ureter. Conversely, excessive analgesia will impair cardiovascular and respiratory function, and may mask local transplant symptoms and signs.

Morphine is the preferred opiate analgesic agent in the initial post-operative period, and can be provided by a continuous intravenous infusion at 1 to 3 mg/hour together with intermittent bolus doses of 2.5 and 5 mg as required. Pethidine is less popular because of the accumulation of metabolites in patients with impaired renal function. Continuous epidural anaesthesia has the disadvantage of requiring careful placement and monitoring, but in the hands of an experienced anaesthetist, is capable of providing excellent analgesia. Fortunately, wound discomfort settles quickly over the first two days, but if continuing analgesia is required, it should be provided by regular intramuscular injections of morphine. The impact of patient controlled analgesia infusions has been to devolve management of pain control to the patient. This has proved effective and thus changed, slightly, the emphasis of analgesia control.

Chemoprophylaxis

It is not unusual to find a renal transplant recipient on multiple medications, some necessary to prevent rejection and others for management of disease in other systems. Nevertheless, despite the concern of polypharmacy and the risk of drug interactions, prophylactic medications may be indicated to reduce the early morbidity of the transplant procedure.

Infection prophylaxis

Transplant wound infection is uncommon (in the order of 1 to 2%), and it is therefore reasonable to restrict prophylaxis to a single dose of a broad spectrum antibiotic at the time of surgery. *Pneumocystis carinii* lung infection is also uncommon, but the severity of infection warrants prophylaxis with a single daily tablet of co-trimoxazole for the first six months. Co-trimoxazole has also been shown to decrease the incidence of urinary tract infections which can occur in about 30% of patients after renal transplantation. Oral acyclovir 200 mg given twice daily has been shown to be effective in reducing the incidence and severity of recurrent oral and genital herpes simplex infection, and in some instances, needs to be given beyond the initial six months after transplantation. The use of immune globulin for the prevention of primary cytomegalovirus infection is discussed in Chapter 13.

Cardiovascular prophylaxis

Pelvic surgery is considered to be a high risk for the development of deep vein thrombosis (DVT). However in the anaemic and uraemic renal transplant recipient, the incidence of a DVT in the first month after surgery is between 1 and 2% and use of heparin unnecessary, particularly when consideration is given to the risk of bleeding after surgery. However, if other risk factors are present such as a high pre-operative haemoglobin of greater than 100 gm/l, prolonged hospitalisation, obesity or a past history of thrombosis, prophylaxis with subcutaneous sodium heparin 5000 units twice or three times daily is indicated. Anti-angina agents that were required before transplantation should be continued afterwards, but their need re-assessed after haemoglobin levels have returned to more normal values.

Gastro-intestinal tract prophylaxis

The widespread use of H_2 receptor antagonists has dramatically reduced the incidence of complicated peptic ulcer disease in the renal transplant population. Indications for use include a past history or symptoms of peptic ulceration at time of transplantation. Ranitidine can be given as a twice daily dose of 150 mg while the corticosteroid dose is at its maximum. Constipation is common in the transplant population but is a subject which is not often discussed. Cyclosporin is thought to affect constipation adversely as does the retroperitoneal transplant surgical procedure. Laxatives may be required in the first weeks of transplantation, with preference for long term treatment given to the intake of a high fibre diet or alternatively, the use of synthetic bulking agents.

Drain tube management

A closed suction drain is usually placed in the extraperitoneal plane to cover possible leakage from the vascular and ureteric anastomoses and divided lymphatics. It is common to drain up to 400 ml of bloodstained fluid in the first 24 hours after transplantation. Large volumes of continuous fresh blood loss, especially in the presence of clinical or ultrasound evidence of peritransplant haematoma, tachycardia or hypotension, is an indication for exploration of the transplant. Haemoserous drainage usually abates within 48 hours, allowing the drain to be removed after the release of suction.

Continuing high volume serous drainage may indicate leakage of either urine or lymph from external iliac lymphatic trunks. They can be differentiated by biochemical analysis with lymph resembling serum, whilst urine has high concentrations of urea, creatinine and often potassium and a low glucose concentration. A dipstick test for glucose can usually distinguish the two alternatives. Urine leakage usually requires early surgical intervention, whilst lymph leakage may be managed conservatively by continued drainage without suction. Irrespective of the volume of lymph being drained, the drain should be removed by at least the sixth day to minimise the risk of local infection, though a lymphocele may develop subsequently (see Chapter 10).

Urinary catheter management

On return to the ward, the urinary catheter and collection bag should be placed in a position such that they cannot be kinked and are clearly visible to the attending staff. The 20 or 22 gauge silastic urinary catheter can usually be safely removed on the morning of the fifth day after transplantation. However, if technical difficulties were encountered with the bladder anastomosis, closure of the bladder or small bladder capacity, it may be advisable to delay catheter removal. Patients with small bladder capacity initially experience urinary frequency which may cause a major problem at night, but these symptoms improve gradually over the following weeks.

Blood in the urine is common for the first few days after transplantation as a result of bleeding from the end of the transplanted ureter, the bladder submucosal tunnel or the anterior cystotomy. The dilutional effect of a high urine output minimises the chance of clot formation. If, however, clots do form and block the catheter, anuria, bladder discomfort, urine bypassing the catheter and bladder rupture may follow. A three way bladder catheter with a continuous bladder washout is not advisable, both because of the dangers that arise from outflow obstruction in this situation and lack of precision in measurement of urine output.

Discomfort related to the presence of the urinary catheter may either manifest as an extreme urge to micturate or as intermittent bladder spasms. If the urinary catheter is not occluded by clot or external pressure, the patient can be reassured that the symptoms will improve quickly. If not, a smaller urinary catheter may help. Antispasmodic agents are usually not of benefit.

A suprapubic catheter may sometimes be necessary at the time of transplantation because of inability to pass the urinary catheter in the presence of urethral stricture or prostatic hypertrophy. It is advisable to delay removal until a successful trial of voiding ten days after transplantation. If the trial of voiding fails, urethroscopy and transurethral surgery will be required.

Fever

Fever in the early weeks after transplantation is common (Table 8.4). In terms of transplant function, it is perhaps best to consider acute rejection

Table 8.4   Recommended antihypertensive agents

Group	Drugs	Comments specific to renal transplantation
Calcium channel blockers	Nifedipine Felodipine Diltiazem Verapamil	* Do not interfere with cyclosporin metabolism * Reduce cyclosporin metabolism requiring cyclosporin dose reduction * Fluid retention
Beta blockers	Atenolol Propranolol Metoprolol	* Detrimental effect on lipid metabolism
ACE inhibitors	Enalapril Captopril	* Monitor for decline in renal function and do not use with non-steroidal anti-inflammatory drugs
Vasodilators/ misc	Prazosin Hydralazine Minoxidil Methyldopa	* Hypertrichosis
Diuretics	Frusemide Metolazone Bumetanide	* Especially in the presence of fluid retention, ACE inhibitors, minoxidil, but not for calcium channel blocker induced ankle oedema

as the likely cause of fever until proven otherwise, particularly if there is evidence of impaired renal transplant function. Investigation of a fever requires culture of blood, urine, sputum and any drainage fluid, chest X-ray, and often, transplant biopsy. On occasions, it may be necessary to commence broad spectrum antibiotics whilst waiting for laboratory results. Presumption that a fever is due to acute rejection and treating without a histological diagnosis is counterproductive if infection is present.

Hypertension

Many factors contribute to the high incidence of hypertension after renal transplantation, including rejection, diseased native kidneys, transplant renal artery stenosis and the effects of immunosuppressive therapy. In the first few hours after transplantation, however, common causes include inadequate analgesia and fluid overload. In the presence of a functioning kidney and a high CVP, fluid restriction or even diuretics may be beneficial. Otherwise, sublingual nifedipine (2.5–5.0 mg), bolus intravenous injections of hydralazine (2.5–5 mg), or a beta blocker can be used to keep the supine blood pressure less than 160/95 mmHg.

More than 60% of transplant recipients remain hypertensive and require regular antihypertensive treatment with commonly used drugs listed in order of preference in Table 8.5. Angiotensin converting enzyme (ACE) inhibitors (e.g. captopril and enalapril) are worthy of particular comment for they are highly effective in reducing blood pressure and

Table 8.5 Common causes of fever in the first month after transplantation

	Signs and symptoms	Investigation	Treatment
Rejection	Graft tenderness Oliguria Raised serum creatinine	Transplant biopsy	Methyl prednisolone OKT3, ALG, ATG
Wound infection	Tender inflamed wound	Wound swab	Drainage & antibiotics
Antilymphocyte preparations	Headaches, nausea, lethargy		Paracetamol, or stop treatment
Septicaemia	Hypotension	Blood culture	Appropriate antibiotic, removal of intravenous lines
Urinary tract infection	Frequency, bladder discomfort	Urinalysis and culture	Antibiotics, removal of catheters or stents
Lung basal atelectasis or pneumonia	Dyspnoea, cough	chest X-ray, sputum culture	chest physiotherapy
Cytomegalovirus infection	Dyspnoea, epigastric discomfort, pancytopaenia, lethargy, anorexia	Serology, blood culture, oesophageal biopsy, bronchial aspirate	Symptomatic or ganciclovir
Peritonitis	Rebound abdominal tenderness	Peritoneal fluid culture	Removal of CAPD catheter, intravenous antibiotics

may help to preserve graft function in the long term by reducing glomerular hyperfiltration. However, ACE inhibitors may precipitate a reduction in graft function especially in combination with non-steroidal anti-inflammatory drugs.

Wound healing

The cellular immune system has an important role in the complex biology of wound healing. Macrophages in particular are responsible for the debridement phase of wound healing and release cytokines that influence angiogenesis, fibroblast proliferation and collagen synthesis. The detrimental effect of corticosteroids on wound healing is well recognised, dose dependent and attributed to the anti-inflammatory action directed against macrophages. There is also evidence in small animal studies to demonstrate that depletion of T cells within ten days of tissue injury results in impaired healing. On the other hand, experimental models have demonstrated no adverse effect of cyclosporin on wound healing.

The reality of current transplant surgical practice is that impaired wound healing and wound infection are not common problems, a claim supported by the difficulty in obtaining data from recent literature on wound infection rates in transplant patients. A study published in 1975, before the introduction of cyclosporin and in the presence of high dose steroids, reported a wound infection rate in kidney transplant recipients of 6.1% which was reduced to 1.6% if infection secondary to either a haematoma or urinary fistula were excluded. The clinical implication is that wounds in immunosuppressed patients will heal provided the technical aspects of surgery performed are of the highest quality. If, however, there is a complication such as a wound infection or a haematoma, particularly in the presence of uraemia, jaundice, malnutrition or diabetes, the immunosuppressed patient does not have the necessary reserve of wound healing capacity to overcome technically poor surgery.

Transplant imaging

The merits of radionuclide and ultrasound examination of the renal transplant are discussed more fully in Chapter 12. Baseline studies in the first week of transplantation are recommended to provide for later comparison, assessing continuity of the collecting system, vascular patency and renal transplant function.

Transplant biopsy

The causes of renal transplant dysfunction are multiple and often multifactorial, and although clinical, biochemical and imaging criteria for rejection are sensitive, they are not specific. Percutaneous biopsy can be performed at any time after renal transplantation, providing a specific histopathological or cytological diagnosis which can be correlated with clinical and other non-invasive criteria to permit rational and safe management of declining renal transplant function (see Chapters 9, 10 and 11). The two types of renal biopsy that are in common use are needle core and fine needle aspiration.

Needle core biopsy

In most instances, the renal transplant can be palpated in the iliac fossa. To reduce the risk of damage to the renal transplant or adjacent small bowel, ultrasound examination should be used either prior to or during the biopsy. The core of cortical tissue is usually be taken from the upper pole, directing the biopsy needle superiorly and laterally depending on the position of the renal transplant (Figure 8.2). Recommended instrumentation for the procedure is an 18 gauge Biopty-Cut biopsy needle mounted in an automated biopsy instrument (Bard Biopty, Covington, GA) producing core biopsies of 20 mm length and 1.2 mm width (Figure 8.3). When fixed in formalin, paraffin embedded sections can be stained using microwave techniques and viewed within three hours of biopsy (Figure 8.4). The procedure can be performed as a day patient admission, with bed rest, observation and examination of urine for a period of four hours after biopsy. Microscopic haematuria is common after biopsy, whereas macroscopic haematuria (occurring in 4%) implies the needle has passed through the intrarenal urothelium and is an indication for admission for observation. The risk of graft loss following 18 gauge percutaneous biopsy is in the order of 1 in 4000. Routine biopsies are advised towards the end of the first week after transplantation to provide baseline histology for subsequent biopsies, and at three months to evaluate the adequacy of immunosuppression.

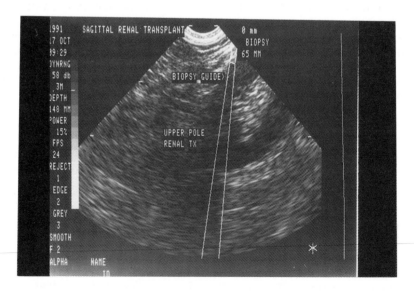

Figure 8.2 With the assistance of ultrasound guidance, the proposed line and depth of percutaneous biopsy of the upper pole cortex are determined.

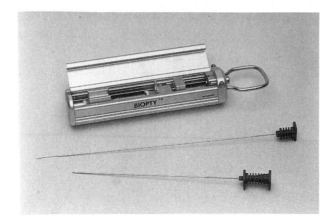

Figure 8.3 Automated 'biopsy gun' and 18 gauge needle used for obtaining a needle core biopsy.

Fine needle aspiration biopsy

In many centres, fine needle aspiration biopsy using a 22 gauge needle has provided safe, reproducible and reliable information on intragraft events. Limitations of the procedure are the need for an experienced cytopathologist and an inability to diagnose vascular or chronic rejection. Differential cell counts are made on the inflammatory cells from both the aspirate taken from the cortex of the renal transplant and a peripheral blood sample. The *in situ* inflammatory component is then determined by subtracting the count of the contaminating peripheral blood from the transplant aspirate (Figures 8.5 and 8.6). The incremental score from each inflammatory cell type is then multiplied by a correction factor based on the importance of that cell in the cellular rejection response (Table 8.6).

Dialysis access

Reliable vascular access is an insurance policy which transplant recipients should be reluctant to part with. Peritoneal dialysis catheters, on the other hand, are an unnatural appendage and an infection risk to be discarded at the earliest opportunity.

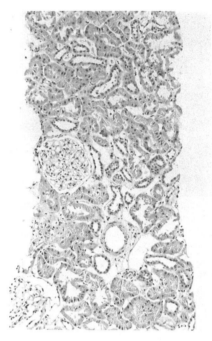

Figure 8.4 Haematoxylin and eosin stained needle core biopsy of transplant renal cortex, obtained one month after transplantation and demonstrating occasional perivascular foci of inflammatory cells (magnification × 108).

Arteriovenous fistulae

Anaemia and impaired clotting associated with endstage renal disease are corrected by successful transplantation and thrombosis of haemodialysis access sites is common. In the early weeks after transplantation, particularly if severe rejection has been encountered, it is advisable to retrieve fistula function by urgent exploration and thrombectomy. Surgical retrieval of the fistula thereafter is less critical and should be performed at the discretion of those responsible for the patient's long term care. Surgical closure of the fistula after transplantation may be undertaken for cosmetic

Table 8.6 Fine needle aspiration cytology incremental scoring system

Cell type	Correction factor
Lymphoblast	1.0
Plasma cell	1.0
Monoblast	1.0
Macrophage	1.0
Activated lymphocyte	0.5
Large granular lymphocyte	0.2
Lymphocytes	0.1
Neutrophil	0.1
Basophil	0.1
Eosinophil	0.1

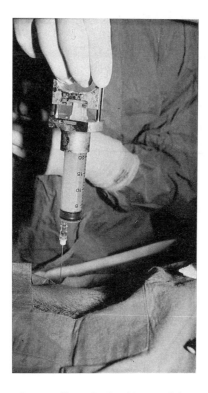

Figure 8.5 Percutaneous fine needle aspiration biopsy of the renal transplant using a 20 ml syringe mounted on a pistol group syringe holder and a 22 gauge needle. Note the small amount of aspirate in the hub of the needle.

reasons or to reduce cardiac output in patients with heart failure. Routine closure of fistulas is not indicated because suitable veins for fistula creation are a limited resource and at least 50% of patients require further treatment for renal failure within ten years of transplantation.

Peritoneal dialysis

Provided the peritoneum is left intact following transplant surgery, peritoneal dialysis can be recommenced as early as the first day after transplantation. In the presence of a leak, however, peritoneal dialysis must be discontinued for at least a week, with alternative haemodialysis provided by either a previously constructed arteriovenous fistula or insertion of a subclavian dialysis catheter. In the presence of a functioning kidney transplant, removal of the peritoneal catheter under general anaesthesia can be undertaken at a time of convenience after the first week of transplantation and can be safely performed without overnight admission.

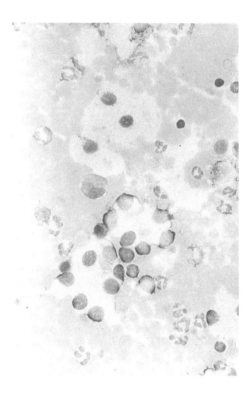

Figure 8.6 Air-dried cytospin preparation of a fine needle aspirate stained with modified Giemsa reagent and demonstrating clumps of renal tubular cells and inflammatory cells (magnification × 400).

Patient education

Most patients receiving a new renal transplant are excited by the possibilities that it offers for improving many features of their lifestyle. Even patients oblivious to their medical care as dialysis patients often become eager to know what they can do to maintain the function of their transplant. By the fifth day after transplantation, uraemia has usually subsided and the lines, drains and catheters removed, and it is therefore an opportune time for medical and nursing staff to commence 'patient education'. Standard booklets can be prepared by the transplant unit to provide essential details about the symptoms and signs of rejection; the nature and importance of their immunosuppressive and other medications; dietary advice with emphasis on low carbohydrate and high dietary fibre intake; and direct lines of communication with transplant staff if problems should arise. Teaching the patient how to measure and record daily the blood pressure, urine output, fluid intake, weight, temperature and urinalysis may prove useful.

It is important to strike a balance between education and transplant awareness on the one hand and development of an obsessive neurosis on the other. This can be achieved by placing emphasis on positive aspects. For example, there is no need to dispose of household pets, nor is there need to avoid crowds of people in enclosed spaces. Sexual activity can be resumed once the recipient has returned home and the discomfort of the transplant procedure has abated. Female patients should take precautions to avoid pregnancy for at least the first year after transplantation, using reliable methods of birth control apart from oral contraceptives which interact with immunosuppressive drugs and increase the risk of thrombosis. It is also advisable for female recipients to void after intercourse so as to reduce the risk of urine infection.

Follow-up after transplantation

The frequency of outpatient attendance after the transplant will be dictated by the time after transplantation and occurrence of problems. The risk of rejection is highest in the first month and dictates the need for daily clinics. Thereafter, attendance is reduced to once weekly by the third month. The end of the third month is a convenient time for routine investigation which may include percutaneous transplant biopsy, ultrasound examination, radionuclide scan and measurement of GFR, together with assessment of lipid levels and glucose tolerance. This assessment signals a change in mode to long term follow-up, in which the patient's very close links with the hospital transplant unit and other health care facilities can be loosened considerably.

9 Primary non-function of the renal transplant

Primary non-function of the renal transplant, also termed initial non-function or delayed function, means different things to different people. Persisting anuria from the time of removing the vascular clamps leaves little ambiguity, but in many cases, urine will flow in small volumes in the early hours after surgery and then decline or stop. A useful working definition of primary non-function is therefore *'the requirement for dialysis within the first seven days after transplantation in a previously dialysis dependent recipient'*. Usually, but not always, this implies that renal preservation has been inadequate, with resultant acute tubular necrosis (ATN). Management of primary non-function is one of the most difficult problems to be encountered in renal transplantation, not only because of the loss of clinical and biochemical parameters to monitor transplant function, but also because the list of important differential diagnoses includes both immunological causes and technical catastrophes (Table 9.1).

The incidence of transplant primary non-function depends upon a number of factors which influence transplant tubular function. The risk of primary non-function increases when the kidney donor and recipient operations are performed in different institutions; with extended warm or cold renal ischaemia times; and when the patient is sensitised or receiving

Table 9.1 Causes of primary non-function of the renal transplant

	Cause	Factors involved
Prerenal	Renal hypoperfusion	Cardiac failure
		Intravascular volume depletion
	Arterial occlusion	Kinked or twisted renal artery
		Proximal stenosis of recipient artery
		Renal artery intimal flap
		Primary thrombosis
		Disruption of anastomosis
Renal	Acute tubular necrosis	Donor and recipient factors (Table 9.3)
		Sequelae of prolonged renal hypoperfusion
	Transplant rejection	Positive crossmatch
		Regraft
		Sensitised recipient
Postrenal	Venous thrombosis	Impaired renal transplant blood flow
	Ureteric obstruction	External compression or kinking of ureter
		Ureteric ischaemia
		Disruption of ureteric anastomosis
	Bladder or catheter obstruction	Clot
		Kinked urinary catheter

a regraft. The rate is lowest (less than 5%) for recipients of living related kidneys, where factors such as donor cardiovascular instability, cold ischaemia time and recipient preparation can be optimised. In recipients of heartbeating cadaver donor kidneys, the reported primary non-function rate for individual units varies between 10 and 50%.

The United Network for Organ Sharing (UNOS) in the USA reported that in the four year period from 1987, 25% of 15,225 first cadaver renal transplants required dialysis in the first week after transplantation. The most significant finding of the analysis was the benefit of primary transplant function, with the one year graft survival being 19% better when dialysis was not necessary (83.2% versus 64.4%). Since donor related factors are dominant causes of both the failure to produce urine immediately and the need for dialysis in the first week after transplantation, there is little doubt that the best approach to the problem is prevention. This chapter, however, considers approaches to the diagnosis and management of inadequate transplant function in the first week after transplantation.

Clinical assesssment

The clinical situation which faces the transplant team on the patient's return from the operating suite is essentially dictated by urine output. If there is a rapid diuresis, management takes a relatively straightforward course centred around careful transplant and cardiorespiratory monitoring as described in Chapter 8. On the other hand, if there is little or no urine output, rapid assessment and effective management may retrieve graft function. In some patients, urine flow may start satisfactorily and then dwindle to nothing over the initial hours after transplantation. Sometimes the drop in urine output will be dramatic, while on other occasions the change will be insidious. There is no substitute for history, examination, measurement and investigation, conducted in that order, by an experienced transplant clinician. Table 9.2 provides a guide for objective clinical assessment of the patient.

History

Many of the answers may be provided by taking a relevant history. A graft that has been removed from a hypotensive or elderly donor, perfused poorly with preservation solutions and then transplanted more than 24 hours later, with long vascular anastomosis times, may be expected not to function immediately. On the other hand, a donor kidney with none of these risk factors transplanted into a highly sensitised recipient, if not functioning immediately, may be suffering from hyperacute or accelerated rejection. Similarly, if urine flow started but then stopped abruptly, perhaps preceded by an increase in haematuria, consideration should be given to catheter blockage or vascular catastrophe.

Pain can be a useful feature of the history. Renal transplantation, like any other operation, is painful to the degree to which analgesia is used for control. The transplanted kidney is denervated and hence pain can be experienced only in the wound or in tissues surrounding the transplant

Table 9.2 Assessment of a patient with primary non-function of the renal transplant

Feature		Implications
History		
Donor history and long ischaemia times		ATN may be expected with poor renal preservation
Urine output	–never started	ATN, vascular problem, hyperacute rejection
	–started then progressive reduction	ATN, accelerated rejection
	–started then stopped suddenly	Vascular catastrophe, blocked catheter/ureter
Pain	–sudden onset over graft	Renal vein thrombosis, graft rupture
	–suprapubic	Blocked urinary catheter, bladder/ureteric leak
Examination		
Poor peripheral perfusion		Assess CVP and cardiac function
Low blood pressure		Assess CVP and cardiac function
Low JVP		Assess CVP
Tachycardia		Assess CVP
Graft tenderness		As for pain, assess drain fluids
Urine haemorrhage		Blocked catheter, renal vein thrombosis
Measurements		
Blood pressure	–low	Blood loss, hypovolaemia, poor cardiac function
	–high	Unlikely to be the cause of non-function
CVP	–low	Check for haemorrhage
	–high	Assess cardiac function
Temperature	–low	Prolonged surgery, high volume fluid replacement
	–high	Rejection
Drain fluid	–blood	Is it fresh, what is the volume?
	–serous	Is it urine or lymph?
Investigations		
Ultrasound	–hydronephrosis	Ureteric (urinary) obstruction
	–reduced or absent blood flow	Arterial or venous thrombosis, ATN, rejection
	–perirenal collection	Urine leak
Drain fluid	–high K$^+$/creatinine/urea, low glucose/Na$^+$	Lymph or blood
	–similar to serum	Seldom helpful, urine osmolality never high, Na$^+$ never low
Urine biochemistry		Changes are more helpful than absolute values
Serum biochemistry		Possible blood loss
Full blood count		ATN, rejection or both
Transplant biopsy		

which can be either stretched or involved in an inflammatory response. After the first 24 hours, the level of pain becomes more discriminatory as wound tenderness declines steadily, and any significant increase in pain from the graft may imply haemorrhage, urinary leak or acute rejection.

Examination

Examination should be directed at the critical problems first. General examination of peripheral perfusion, arterial blood pressure and intravascular volume from the jugular venous pressure (JVP) and central venous pressure (CVP) help to elucidate many cases of poor transplant function. As a general rule, the state of kidney perfusion will be mirrored by perfusion of the feet, unless of course there is specific peripheral vascular disease. If the blood pressure is low and associated with poor perfusion and a high JVP, urgent attention to cardiac function will be needed. On the other hand, if the intravascular volume is low the challenge is to fill the patient with colloid solutions and blood to rectify the situation. If the patient is hypothermic, warming blankets are required.

Abdominal examination must include gentle palpation of the transplant and surrounding tissues, percussion of the bladder size and auscultation for renal bruits. Serial examination of the transplant provides more diagnostic help than a single examination.

Two tubes need careful consideration. Firstly, the urinary catheter must be examined for haematuria, clot and obstruction. Most nursing staff and clinicians will at some time have the opportunity to diagnose and treat 'anuria' by the simple action of untwisting the catheter, removing it from under a patient's thigh, or untangling it from the patient's underwear. Free flow of urine back down the catheter into the bladder under gravity is the easiest way of establishing whether or not it is blocked at the tip. Failing this, diagnostic bladder washout with 30 ml of sterile water or saline without force should be regarded as part of the examination. Secondly, there may be a wound drain, down which can be flowing blood, urine, lymph, or small quantities of serosanguinous discharge. Only the last of these scenarios may be accepted with equanimity. All the other possibilities should be further investigated by measuring electrolytes and haematocrit in the drain fluid.

Investigation

Ultrasound examination has become almost an extension of the clinical examination in many fields of medicine. This is true in transplantation when, if the diagnosis is not already obvious, ultrasonography may provide bedside diagnosis without the need for invasive and time consuming studies in the radiology department. Importantly, ultrasound examination is able to demonstate blood flow into and out of the transplant as well as continuity of the urinary collecting system. If doubt remains, the differential diagnosis will usually be between ATN and rejection. The choice is then straightforward, for one can either treat on the balance of probability or perform an early percutaneous biopsy of the kidney and treat on the basis of histology.

Renal transplant hypoperfusion (prerenal failure)

The renal transplant weighs approximately 150 gm and has a blood flow of about 600 ml/min, which represents up to 20% of the resting cardiac output. Glomerular filtration ceases if the afferent glomerular arterial mean pressure falls below 60 mmHg. Therefore, renal transplant blood flow and function are susceptible to impaired cardiac output either because of a low intravascular volume or cardiac failure. Both these situations are frequently encountered in the patient presenting for transplantation with intravascular fluid depletion common both in the first 24 hours after a haemodialysis and in peritoneal dialysis treated patients. Myocardial function can be impaired by coronary artery disease and the presence of uraemia. Fluid management of the renal transplant recipient, from the time of presenting to hospital for the transplantation procedure, must therefore be designed to produce maximal perfusion of the renal transplant by minimising the feedback effect on arteriolar smooth muscle produced by an inadequate circulating volume (see Chapter 8).

Prerenal failure may be reversible after effective restoration of the intravascular volume and body temperature. Marked or prolonged reduction in transplant blood flow, however, will exacerbate ischaemic damage to the renal tubules, compound the preservation injury and produce acute renal failure. Rapid correction of renal hypoperfusion is thus important, with replacement of blood loss and maintenance of the CVP as high as 15 cm H_2O. Dopamine, when used in doses of 2 to 4 μg/kg/min, will assist glomerular filtration, but diuretics are rarely of benefit. In the presence of persisting systemic hypotension, despite a high CVP, an electrocardiograph may detect acute myocardial ischaemia or infarction. Inotropic agents such as dobutamine may be needed to support the circulation in these circumstances.

The decision to restrict further hydration and accept that tubular cell damage is present, is often finely balanced and requires clinical skill. Too much fluid replacement will precipitate pulmonary oedema leading to emergency ultrafiltration or haemodialysis, whereas inadequate replacement compromises renal transplant function.

Vascular catastrophe

Impaired or absent renal artery blood flow is an uncommon cause of transplant non-function, occurring in perhaps one to 3% of procedures. Unless identified during surgery, the effects of arterial occlusion are generally irreversible by the time the diagnosis has been suspected, let alone confirmed by ultrasound examination, angiography or a radionuclide scan (Figure 9.1).

Arterial occlusion

At the time of transplant surgery, poor perfusion of the renal transplant may follow hyperacute rejection, spasm of the transplant renal artery, malrotation of one or other of the vascular anastomoses, or an intimal flap

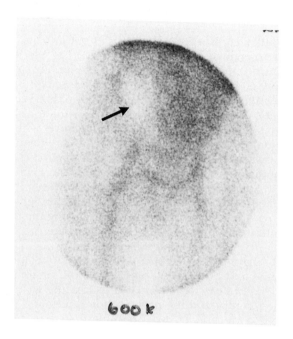

Figure 9.1   DTPA radionuclide scan demonstrating an area with no perfusion in the right iliac fossa (arrow) in a 60 year old recipient with a right renal transplant. The investigation was prompted by sudden onset of painless anuria three days after transplantation. The transplant renal artery was occluded by an embolus from the left ventricle of the heart.

produced by traction on the renal artery. While spasm resolves spontaneously, both unrecognised malrotation and intimal flaps will subsequently result in the thrombosis of the renal artery.

Even though it is often difficult to position the renal transplant so as to ensure uninterrupted arterial and venous blood flow at the completion of the vascular anastomoses, particularly for right sided donor kidneys and those that have multiple arteries or early arterial branching, subsequent arterial thrombosis is unusual. Paediatric donor kidneys, however, irrespective of the age of the recipient, are more likely to thrombose, perhaps because of the smaller arterial lumen or the increased mobility and propensity for kinking of the vascular pedicle.

Primary arterial thrombosis may occur without a mechanical cause in the presence of low renal transplant blood flow. Systemic hypotension, dehydration or ATN causing increased intrarenal pressure may contribute. The terminal event for a poorly preserved donor kidney that never functions is usually arterial thrombosis. Equally, arterial thrombosis may be the terminal event for irreversible acute rejection. Cyclosporin has been suggested to be a causal factor for thrombosis; however there is no definitive evidence to support this relationship.

Acute transplant renal artery occlusion is painless and frequently discovered at time of ultrasound or radionuclide examination to investigate the absence of transplant renal function. By this time, the renal transplant parenchyma will usually have been irreversibly damaged by warm ischaemia. Nevertheless, immediate exploration of the transplanted kidney in the operating suite is indicated. Recently transplanted kidneys are easily mobilised and will be small and dark purple in appearance. If a pulse is still present in the renal artery, a small incision can be made in the cortex to observe for arterial bleeding. If there is still doubt regarding the viability of the renal transplant, a cortical biopsy can be taken and a frozen section examined. After the kidney has been removed, it should be subjected to a macroscopic and microscopic examination by a histopathologist with attention given to the possibilities of arterial injury or rejection.

Late presentation of arterial occlusion, several hours after the event, is associated with increasing discomfort and tenderness in the region of the renal transplant as an inflammatory response is mounted in tissues adjacent to the infarcted renal transplant.

Venous thrombosis

The causes of transplant renal vein thrombosis, which occurs in up to 5% of transplanted kidneys, are unclear. However, as described in detail in Chapter 10, the usual presentation with graft rupture is much more dramatic than that of arterial occlusion.

Graft rupture

Rupture of the kidney occurs in about 1% of renal transplants and can be a life threatening event occurring usually between four and 14 days after transplantation. The rupture occurs along the convex border of the renal transplant between the upper and lower poles as a result of increased intrarenal pressure. At the time of rupture, an adult renal transplant measured by ultrasound may be up to 16 or more centimetres in length. The principal cause of the increased intrarenal pressure has changed over the last decade. ATN following organ donation from non-heartbeating cadaver donors was the commonest cause of rupture in the early years of renal transplantation and often prompted transplant surgeons to perform a prophylactic capsulotomy at transplantation. More recently, with the advent of heartbeating cadaver donors, renal vein thrombosis has become the more likely cause.

Rupture of the renal transplant causes extreme discomfort around the renal transplant. The signs are a tender mass and hypotension. Immediate exploration of the renal transplant is recommended with evacuation of the surrounding clot and examination of the renal vessels, since it may be possible to remove the thrombus from the transplant renal vein. If the renal vessels are patent, attempts can be made to control cortical bleeding by diathermy, local pressure or application of synthetic aids to coagulation. In the presence of ATN, release of intrarenal pressure by the rupture may alone be sufficient to control cortical bleeding. A biopsy of the renal transplant should be undertaken under direct vision to exclude acute rejection.

Acute tubular necrosis

Most patients with initial non-function after renal transplantation have been transplanted with a donor kidney damaged by the effects of ischaemia following imperfect preservation between the time of donor surgery and completion of the transplant vascular anastomoses. ATN is a term used to describe the functional cellular injury, even though frankly necrotic tubular epithelial cells may not be demonstrated on light microscopy examination of a needle core biopsy. Management of the renal transplant with ATN, because of the need to continue dialysis, the lack of simple biochemical markers of rejection and the need to avoid nephrotoxic drugs, is inconvenient at the very least.

Clinical presentation of ATN

The presentation of ATN varies from subtle functional defects that do not affect urine volume to gross tubular cell injury and anuria present for up to three months after transplantation. Most present between these two extremes, with some patients avoiding the need for dialysis and others requiring dialysis for a week or two. The rate of recovery of tubular function will depend upon the extent of the initial injury and the presence of nephrotoxic agents such as cyclosporin and aminoglycoside antibiotics. In the context of primary non-function, the tubular ischaemic injury occurs at the time of donor kidney preservation or soon after trans- plantation. However, the same tubular injury can occur in a functioning renal transplant at any time after transplantation.

Pathogenesis of ATN

There are many risk factors for the development of ATN (Table 9.3), but the common endpoint is cellular damage as a result of a supply of oxygen inadequate to meet the metabolic requirements of the kidney at that time, whether warm and perfused in the donor, stored in ice, or revascularised in the transplant recipient. The pathogenesis of ATN is best understood

Table 9.3   Risk factors for development of acute tubular necrosis in renal transplants

1.	Donor hypotension
2.	Non-heartbeating cadaver organ donation
3.	Cerebrovascular accident as cause of donor death
4.	Donor age greater than 50 years or less than ten years
5.	Prolonged warm ischaemia time of greater than five minutes between cross clamping aorta and commencement of cool perfusion with organ preservation fluid
6.	Inadequate renal perfusion with organ preservation fluid
7.	Prolonged cold storage of greater than 24 hours, but more so for greater than 48 hours
8.	Long vascular anastomosis time of greater than 40 minutes
9.	Recipient hypotension or hypovolaemia after transplant revascularisation

at a cellular level, where the primary event is a decrease in aerobic energy metabolism. Paralysis of the energy dependent cell membrane sodium/potassium pump follows, and leads to loss of intracellular electrolyte homeostasis. Extracellular sodium and chloride, together with water, pass into the tubular cell along concentration gradients, to produce cellular swelling. Cytoplasmic calcium concentration, which increases as a result of redistribution of calcium within the cells, activates autolytic enzymes.

Fortunately for the purposes of transplantation, the energy requirements of the cell are temperature dependent. Anaerobic metabolism can maintain cellular energy requirements for periods of up to 48 hours provided the cell has been cooled to about 4°C with appropriate preservation fluids (see Chapter 3). In contrast, ischaemia at 37° can be tolerated for about five minutes, after which progressive cellular injury occurs and renders the kidney non-viable within 30 minutes.

Oxygen and metabolic substrate delivery return to normal after revascularisation of the transplanted kidney. However, tubular cell injury may continue because of the formation of oxygen free radicals and increased intracellular phospholipase activity, leading to both plasma membrane and intracellular membrane damage with destruction of the cell. This process may in part explain the commonly encountered situation where there is early urine production followed by oliguria within hours of transplantation. The pathogenesis of the reduced glomerular filtration rate with accompanying oliguria or anuria is nevertheless controversial and probably multifactorial. Blood flow to the renal cortex is reduced in most cases of established ATN as a result of increased intrarenal pressure from intracellular and extracellular oedema confined by the renal capsule. As the necrotic tubular cells are replaced by regenerating cells and oedema resolves, renal blood flow improves and renal transplant function returns.

Investigation of ATN

Failure of the renal transplant to function within one or two hours, despite adequate hydration of the recipient, is usually diagnostic of ATN. The purpose of any further investigation is to exclude rejection, to assess the severity of the tubular damage and to monitor the processes of tubular repair. In the presence of raised intrarenal pressure both ultrasonography and radionuclide scanning will demonstrate impaired cortical blood flow but do not exclude a diagnosis of acute rejection (see Chapter 12). As the intrarenal pressure decreases, renal blood flow improves, and eventually there is evidence of urinary excretion of radionuclide tracer material. The definitive investigation is a needle core biopsy examined by light microscopy. Until such time as renal function returns, regular ultrasonography and renal biopsy should be performed at least at weekly intervals.

Histopathology of ATN

The characteristic light microscopy findings are tubular dilatation, the absence of individual tubular cells, loss of tubular cell nuclei, and cell to cell variation in size and shape (Figure 9.2). Flattened cells, often with extruded nuclei, represent damaged cells. Larger and sometimes

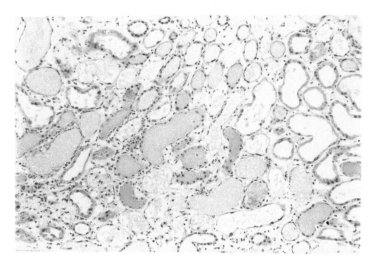

Figure 9.2 Renal cortex biopsy taken eight days after transplantation, demonstrating dilated tubules with flattened tubular cells, loss of tubular cell nuclei and absence of individual tubular cells (magnification × 82)

multinuclear cells, often with an increase in mitotic figures, represent regenerating tubular cells. Interstitial oedema with an increase in space between the tubules is a prominent finding and correlates well with tubular dilation. Polymorphs are frequently identified and are involved in clearing cellular debris. The glomeruli often show no significant change by light microscopy; however on electronmicroscopy, flattening and spreading of podocytes can be identified.

Although there is a clear association between severe structural changes and a low glomerular filtration rate, it is common to find marked impairment of renal function accompanying minimal structural change. Conversely, marked structural changes can sometimes be associated with near normal renal function.

Management of ATN

Prevention is definitely better than the cure because, following confirmation of the diagnosis of established ATN and exclusion of other causes of primary transplant non-function, there are no factors that are known to have a beneficial therapeutic effect. Before ATN has become established, however, urgent attention to the patient's fluid status and infusion of dopamine probably rescues function in a proportion of grafts. If these measures have failed, attention must turn to minimising the duration of ATN, which may last for up to three months, and to diagnosing intercurrent rejection episodes. This can be a test of patience for the transplant recipient who often needs to be reassured of the probable satisfactory outcome.

Cyclosporin is nephrotoxic and is known to prolong ATN. Many transplant centres thus attempt to ensure graft function has been established before using cyclosporin, adopting either an 'induction strategy' with polyclonal or monoclonal antilymphocyte preparations, azathioprine and prednisolone or less frequently, relying on azathioprine and prednisolone alone. In the presence of persisting ATN, the anti-lymphocyte preparations can be continued for ten to 14 days, after which it is wise to reintroduce cyclosporin at dose of 4 to 6 mg/kg/day, even though there may be little or no evidence of transplant renal function. In this situation, prevention of rejection is more important in the first month than the presence of ATN. Thereafter, the cyclosporin dose may be more safely reduced or withdrawn in an attempt to promote transplant function. It is of course also important to minimise use of nephrotoxic agents such as the aminoglycosides in this early and fragile phase (see Chapter 10 and Appendix 3). There may also be a theoretical advantage to the administration of prophylactic subcutaneous heparin to reduce the risk of vascular thrombosis in the poorly perfused transplant with ATN.

Dialysis treatment will be needed unless the transplant functions within 48 or 72 hours. It is common practice to aim for a body weight after dialysis of 1 or 2 kg in excess of ideal dry body weight before transplantation, in order to prevent dehydration and improve graft perfusion. While some patients tolerate a greater fluid load, it would be unsafe to use OKT3, if subsequently indicated, if the body weight is more than 3 kg above the ideal level. Providing that dialysis can be continued on an outpatient basis, there is usually no reason to continue hospitalisation of the recipient with ATN after the initial seven to ten days.

The major management challenge is to detect allograft rejection at an early stage when deprived of urine output and serum creatinine as effective indicators of transplant function. Careful and regular clinical examination may reveal graft swelling or increasing tenderness. Rising temperature should alert one to the possibility of rejection, rather than assuming chest or urinary tract infection. Serial surveillance with ultrasound Doppler or radionuclide assessment of renal blood flow together with at least once weekly transplant biopsy are appropriate.

Evidence of recovery of transplant renal function will be a slow increase in the daily urine volume followed by clearance of urea and creatinine. In some instances, polyuria will result from the loss of ability to concentrate the urine and attention to fluid and salt replacement is needed. Some patients may need intravenous hydration to maintain their fluid replacement, especially if it occurs in the first few days after transplantation.

Hyperacute rejection

Intrarenal causes of primary transplant non-function also include immunologic mechanisms. Hyperacute rejection may follow binding of circulating donor specific antibodies to donor vascular endothelial antigens. This can occur after transplantation across ABO blood group barriers or when the recipient has been sensitised against donor histocompatibility antigens as a result of previous transplants, transfusions or pregnancies. Crossmatching of donor lymphocytes and recipient serum prior to transplantation has ensured that hyperacute rejection is now uncommon (see Chapter 5). However, a variant of the classical appearance of rapid and irreversible destruction of the renal transplant following revascularisation may explain the increased rate of non-function in recipients of regrafts.

Clinical presentation

The surgeon will have observed that the transplant, instead of reperfusing evenly, turning pink and becoming firm with pulsatile hilar blood vessels after removal of the vascular clamps, has remained flaccid and blue or patchy. In some cases, the kidney may have perfused well initially, but then become pale or patchy in appearance within an hour or so of completing the vascular anastomoses. A biopsy taken at this time may show changes on light microscopy ranging from accumulation of neutrophil polymorphs in engorged glomeruli through to a dramatic picture of destruction. When fully developed, hyperacute rejection leads to thrombosed glomeruli, fibrinoid necrosis and thrombosis of arterioles, interstitial haemorrhage and cortical necrosis (Figure 9.3).

Pathogenesis

The mechanism of hyperacute graft destruction involves preformed circulating antibodies in the recipient binding to antigens on the donor vascular endothelium. Complement binding damages the endothelial cells and leads to recruitment of effector cells such as polymorphs and macrophages. Platelet aggregation on the damaged endothelium triggers vasoconstriction, further local damage and thrombosis. In many cases, but not all, it is possible to define the causative antibody which may be directed at donor ABO antigens, HLA antigens or other polymorphic antigens restricted to the vascular endothelium. Prediction of this interaction and prevention by transplanting the kidney into someone else is clearly the best course.

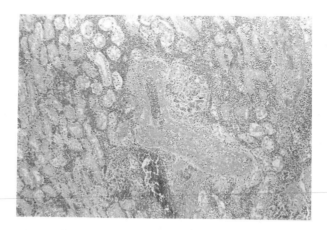

Figure 9.3 Section of renal cortex from a transplant nephrectomy specimen 24 hours after transplantation demonstrating arterial fibrinoid necrosis, thrombosis of arterioles, interstitial haemorrhage and cortical necrosis (magnification × 80)

Management

Treatment of hyperacute rejection by removal of circulating antibodies by plasmapheresis has been used with isolated success. Corticosteroids, antilymphocyte preparations and baseline levels of immunosuppression, while necessary, probably have little influence on the outcome of hyperacute rejection. Definitive management is usually by nephrectomy, either at the transplant operation if graft destruction is apparent, or more commonly, when blood flow to the transplant ceases in the first six to 24 hours after transplantation.

Accelerated rejection

Severe and florid rejection can be encountered one to four days after transplantation. Clinical presentation may include the development of oliguria over a matter of hours in a previously well functioning transplant, high fever of more than 38.5°C and marked discomfort and tenderness in the region of the transplant. The timing of the event is too early for classic cell mediated acute rejection and implies previous exposure by the recipient to antigens present on the renal transplant. The rejection process occurs in the absence of preformed circulating antibodies and despite a negative crossmatch with donor lymphocytes. However, it is probable that specifically sensitised host T and B cells are capable of mounting this accelerated rejection episode. This hypothesis is supported by histology findings of a mixed cellular and humoral rejection episode with both interstitial cellular infiltration and severe vascular damage with necrotising vasculitis. Once the diagnosis of accelerated rejection is suspected, immediate intervention is warranted and should include a needle core biopsy and intravenous methylprednisolone before the biopsy result is available. The rejection process may be difficult to control even with the use of antilymphocyte preparations.

Ureteric and bladder complications

In the catheterised renal transplant recipient, obstruction can occur anywhere between the renal calices and the urine collection bag. Most cases are readily reversible being due either to a technical problem with kinking of the ureter, obstruction of the ureter at the site of ureteroneocystostomy or obstruction of the urinary catheter by clot in the bladder or catheter. The effects of ischaemia at the lower end of the ureter do not usually become apparent until four or five days after transplantation (see Chapter 10). In all unexplained cases of transplant primary nonfunction, it is important to rule out urinary catheter obstruction by performing a bladder washout on return to the ward. Ultrasound examination is capable of detecting a dilated collecting system and will identify a blood clot in the bladder. Return to the operating theatre is needed for ureteric obstruction or if large amounts of clot are encountered in the bladder.

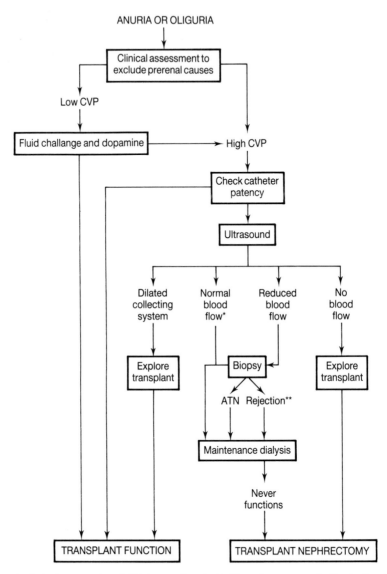

Figure 9.4   Integrated plan for the management of primary non-function of the renal transplant.

Integrated management plan

The most likely cause of primary non-function of the renal transplant is ATN, and as emphasised in this chapter, management revolves around the need to exclude other causes and maintain surveillance for rejection. Figure 9.4 demonstrates the two eventual outcomes of management, either transplant function or the need for graft nephrectomy of a failed transplant. The challenge is clearly to avoid the latter.

Further reading

1.   Cecka JM, Cho YW, Terasaki PI. Analysis of the UNOS scientific renal transplant registry at three years – early events affecting transplant success. *Transplantation* 1992; **53**, 59–64.

10   Early renal transplant dysfunction

The first three months after transplantation are the most critical in terms of survival and normal function of both the transplanted kidney and the patient. This period is also the most rewarding and challenging for the members of the transplant team, as they witness the rejuvenation and rehabilitation of the recipient. Successful transplantation is, however, not achieved easily, with perhaps as few as a quarter of kidney transplant recipients experiencing an uncomplicated course through the first year. For the majority, troublesome renal transplant dysfunction can occur from one or more causes summarised in Table 10.1. Success also acts as a salutary reminder of failure, particularly when associated with poor management decisions and missed therapeutic opportunities.

The survival curve after kidney transplantation, as with other forms of solid organ transplantation, falls most rapidly in the first three months after transplantation when between 10 and 20% of transplants are lost. Acute allograft rejection has traditionally been the commonest cause of graft loss during this period, but over the last decade, with use of more effective immunosuppression protocols, the trend has been for a gradual reduction in graft loss as a result of rejection (Figure 10.1). Hence, increased emphasis must now be placed again on the need to avoid graft loss from causes related to surgical technique or patient death.

The spectrum of the clinical presentation of deterioration in transplant function is considerable, varying from abrupt and life threatening onset of anuria with dramatic local clinical signs to a gradual decline in renal function as evidenced by an insidious increase in serum creatinine and

Table 10.1 Causes of early renal transplant dysfunction

Prerenal
1. Dehydration
2. Hypotension – cardiogenic shock
                 – salt depletion
                 – bacteraemic shock
3. Arterial   – thrombosis
                 – embolic
4. Venous   – thrombosis
5. Death with functioning transplant

Renal
1. Rejection
2. Acute renal failure (ATN)
3. Drug nephrotoxicity
4. Recurrent primary renal disease
5. Infection   – CMV
                       – pyelonephritis

Postrenal
1. Ureteric   – clot
               – kink
               – external compression
2. Bladder   – clot retention
               – neuropathy
3. Prostatomegaly and bladder neck obstruction
4. Urethral stricture

---

no local signs. In the presence of acute tubular necrosis (ATN), awareness of additional causes which compromise transplant function requires
increased clinical surveillance together with frequent use of transplant
imaging techniques and histopathological assessment. In this chapter,
emphasis is placed on obtaining an accurate and early diagnosis of the
cause of deterioration in renal transplant function instituting appropriate
measures to minimise irreversible loss of nephrons and maximising patient
survival.

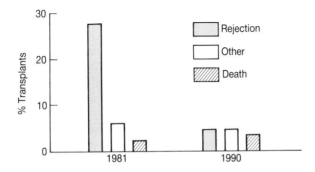

Figure 10.1 Comparison of percentage graft loss in the first six months after transplantation,
in Australia, in calendar years 1981 and 1990, as a result of rejection, non-immunological causes
(other) and death with a functioning graft (ANZDATA Registry Report 1991).

Acute rejection

The human immune system is capable of distinguishing self from non-self and provides the ability to mount a defence against viruses, bacteria, mutant and other foreign cells. Unless the cells of the transplanted kidney have the same surface antigens as the recipient, T lymphocytes will respond to the graft leading to a complex chain of cellular and antibody mediated responses that will result in graft cell destruction and eventual vascular thrombosis. Clinical evidence of this classical 'first set' rejection response by the immune system is usually not encountered until at least five days after transplantation, and is heralded by a deterioration in kidney transplant function, local signs and distinctive histopathological changes.

Evidence of an immune response within the first five days after transplantation implies that the transplant recipient's immune system has been exposed previously to at least one of the cell surface antigens. This exposure may have occurred at time of blood transfusion, previous transplantation or as a result of pregnancy. The earliest form of rejection, the so-called hyperacute rejection, may occur if pre-existing anti-donor antibodies are circulating at the time of transplantation of the kidney, with dramatic and sudden loss of graft function within minutes to hours after transplantation. The primary target of these antibodies are antigens on the vascular endothelium of the transplanted kidney (see Chapter 9). Organs transplanted into ABO blood group incompatible recipients and xenografts are lost in a similar manner.

Evidence of transplant rejection between two and four days after transplantation is considered a 'second set' response, when cells of the transplant recipient immune system have been specifically sensitised by previous exposure to one or more of the transplanted kidney antigens. This 'accelerated' rejection response includes both cellular and antibody components, although vascular damage is often predominant and is the probable explanation for the abrupt deterioration in graft function seen in some living related kidney recipients who received donor specific blood transfusions prior to transplantation.

By convention, acute rejection is described as the immunological response to the transplanted kidney that may become evident as early as the end of the first week after transplantation and usually occurs within the first three months after transplantation. Chronic rejection is usually responsible for slow deterioration of renal function generally beyond the first three months after transplantation. The result of chronic rejection is progressive atrophy of nephrons which are, in turn, replaced by fibrous tissue. For a more comprehensive appraisal of the immune response and allograft rejection, the reader is referred to texts listed at the conclusion of this chapter.

Clinical presentation

The classic symptoms of acute rejection are fever, discomfort in the region of the renal transplant, oliguria and malaise. On examination, the transplant recipient is frequently hypertensive, with a tachycardia and temperature between 38 and 39°C. The denervated renal transplant is tender because of irritation of the adjacent peritoneum by the immune inflammatory response within the renal transplant. Progressive enlargement of the transplant results from increasing oedema within the graft and is accentuated by oedema in recipient tissues around the graft. This classic presentation, towards the end of the first week after transplantation, is almost diagnostic of acute rejection, but has been seen less often since the introduction of cyclosporin immunosuppression.

In the presence of cyclosporin, the symptoms and signs of acute rejection are less dramatic, with temperatures frequently less than 38°C. There is little or no transplant tenderness and less prominent transplant enlargement. Equally, the decline in urine output is usually less abrupt. Subtle changes of worsening hypertension and weight gain may be observed as a result of decreased sodium urinary excretion. The clinical diagnosis of acute rejection has therefore become difficult to make, particularly when the presence of anuria or oliguria during the recovery phase of co-existing ATN deprives the clinician of a simple measure of renal function. These changes, induced by the need for improved immunosuppression provided by cyclosporin, have therefore placed greater emphasis on the need for additional information from diagnostic imaging and histopathological assessment.

Diagnostic investigations

In a functioning renal transplant, the serum creatinine level is an easy, early, cheap, but not necessarily specific, marker of renal transplant rejection. An increase, or failure to observe a continuing fall, in the serum creatinine level may be indicative of rejection. Serum urea levels are less reliable because of fluctuations associated with diet, state of hydration and use of high doses of corticosteroids for the treatment of rejection. Leucocytosis is a frequent accompaniment of acute rejection but is also non-specific and less prominent in the presence of cyclosporin treatment. The morphology and differential count of peripheral white blood cells is unhelpful, with the types of white cells present not being representative of those present in a rejecting transplant.

Radiologic imaging of the rejecting renal transplant is of value, but can neither confirm nor establish a diagnosis of rejection (see Chapter 12). Ultrasonography is the most helpful, particularly if frequent sequential assessment is undertaken. An increase in the size of the graft and a reduction in the amplitude of arterial flow as assessed by duplex scanning may be indicative of rejection, but can also be seen with ATN and obstruction of the urinary collecting system. Perhaps the most important result of ultrasound assessment is exclusion of non-immmunological causes of renal dysfunction such as urinary collecting system obstruction or vascular thrombosis. Isotope scanning of the renal transplant is sensitive in demonstrating abnormalities in uptake of the isotope tracer by the renal tubular cells and excretion into the urine. However, reduction of tracer uptake is non-specific and is also seen in ATN, cyclosporin nephrotoxicity, pyelonephritis and other parenchymal diseases. Raised intrarenal pressure, measured by an ultrasound guided needle placed in the renal cortex and connected to a manometer, may be indicative of rejection, but can also be seen with ATN.

The gold standard used by most clinicians to confirm a diagnosis of rejection is the histopathological assessment of a needle core biopsy of the renal transplant cortex. The percutaneous biopsy technique described in Chapter 8 has a low morbidity and can be assessed by light microscopy within three hours. However, even the biopsy is open to different interpretations and should be evaluated in the context of the patient's clinical presentation. For this reason, it is important for transplant clinicians to review the light microscopy slide preparation, since it is they who must make the decision and take on the responsibility of treatment of an episode of acute rejection. Weekly review of the transplant unit's histopathology specimens is usually instructive for all members of the transplant team and offers positive feedback for the committed transplant histopathologist.

Morphologic examination of fine needle aspirates from the renal transplant in transplant centres with the appropriate expertise of a cytopathologist is also a proven technique for monitoring events within the transplant, particularly cellular rejection. It is unreliable for the diagnosis of vascular rejection where anatomical information is required.

Histopathology

Infiltration of mononuclear cells into the interstitium between the renal tubules is regarded as the most important and characteristic lesion of acute renal transplant rejection. The degree of infiltration is, however, not necessarily related to the severity of the rejection response and can occur either in the presence or absence of clinical rejection. The infiltrate is comprised predominantly of T lymphocytes together with monocytes and plasma cells. The interstitial mononuclear infiltrate may be diffusely distributed in the biopsy and associated with varying degrees of interstitial oedema (Figure 10.2). These cells usually migrate between the adjacent endothelial cells in reponse to locally released migration factors, without causing serious damage to the wall. When the capillary wall is damaged, there may be an associated interstitial haemorrhage which is indicative of severe and often irreversible rejection. Although the initial mononuclear infiltrate seen in the renal transplant results from migration of recipient cells, the majority of mononuclear cells seen within the renal transplant, with the T cells or B cells, are probably the result of cellular division within the renal transplant. Focal interstitial cell infiltrate, principally around small veins, is a common finding usually associated with no clinical or laboratory evidence of rejection (Figure 10.3). Apart from occasional areas of tubular invasion by lymphocytes, changes to the tubular epithelium are often unremarkable in early acute rejection, despite the presence of prominent infiltration by mononuclear cells. In the more severely affected renal transplants, the tubules undergo ischaemic changes as a result of vascular damage, leading to the histological appearance of ATN.

Vascular and glomerular changes are presumed to be the consequence of antibody mediated rejection, and are frequently seen in conjunction with cellular rejection. Platelet aggregates, often occluding the vessel, can be seen in glomerular and peritubular capillaries. Small arteries and arterioles exhibit intimal changes of oedema and infiltration of mononuclear cells. As the immune process becomes more severe, the walls of the arterioles

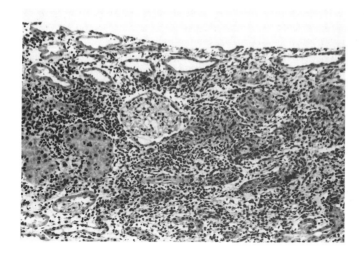

Figure 10.2 Renal cortex biopsy six days after transplantation demonstrating severe diffuse mononuclear cell infiltrate, invading renal tubules and interstitial oedema consistent with acute rejection (H&E,x50).

become homogeneous in appearance and are described as having fibrinoid necrosis, usually associated with irreversible damage to the renal transplant (Figure 10.4). Similar changes are seen in the larger arcuate and hilar arteries. Cortical necrosis occurs as capillary and small vessel thrombosis becomes more widespread. Even though blood flow to the renal transplant may continue, and the renal medulla may remain viable, the biopsy finding of cortical necrosis represents endstage rejection and, unless there is

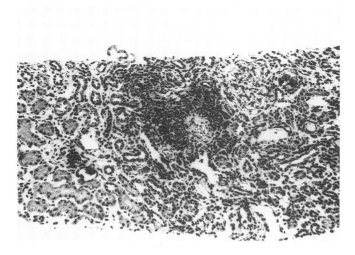

Figure 10.3 Renal cortex biopsy three months after transplantation demonstrating focal perivascular mononuclear cell infiltrate. Some arteriolar intimal thickening is present, perhaps consistent with previous vascular rejection (H&E,x40).

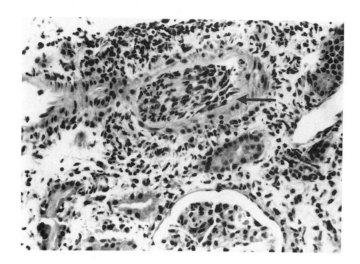

Figure 10.4 Renal cortex biopsy five days after transplantation demonstrating mononuclear cell infiltrate within the swollen intima of an interlobular artery (arrowed) with resultant occlusion of the lumen (H&E,x80).

evidence that the process is patchy, is an indication for removal of the renal transplant (Figure 10.5).

Treatment of acute rejection

The first step in the management of any episode of rejection is to make a confident and early clinical diagnosis, which should be supported at all times by a histological diagnosis. Depending on the severity of the rejection response, any delay in diagnosis and initiation of treatment may result in an irreversible loss of nephrons, if not the whole renal transplant. While the transplant clinician should be aggressive in the search for evidence of rejection, it is important to be both careful and circumspect in the choice of treatment, since patient safety and survival is of far greater importance than renal transplant survival. The implication of the diagnosis, and the rationale for treatment of acute rejection, is that the transplant recipient is inadequately immunosuppressed. However, if there is clinical evidence of profound immunosuppression from indicators such as pancytopaenia and opportunistic infections, additional immunosuppression is probably unwarranted, irrespective of the appearance of the transplant biopsy.

Conventional treatment of acute rejection involves the use of intravenous bolus injections of methylprednisolone for between three and five days at a dose between 250 and 1000 mg depending on patient size and subjective assessment of the severity of the rejection episode. Evidence to support one dose over the other is not forthcoming. Isolated reports of cardiac arrest during rapid bolus injection of steroid in transplant patients has prompted the recommendation that methylprednisolone be given as an infusion over at least five to ten minutes. The use of intravenous methylprednisolone is effective in about 80% of cases of acute cellular rejection and, compared to the alternative antilymphocyte preparations, is of low cost (see Chapter 7). Resolution of the fever and marked improvement in the local signs may be seen within an hour of steroid injection, while the serum creatinine falls over the ensuing 24 hours. A rapid clinical response is usually associated with a favourable prognosis. During the course of treatment, the peripheral white cell count may increase further, serum urea remain disproportionately raised, and blood cyclosporin levels become elevated as a result of competition of liver metabolism by the cytochrome P450 system. It is wise to monitor blood glucose levels which rise significantly in up to 10% of non-diabetic recipients.

Although some transplant centres prefer to use either polyclonal (ATG/ALG) or monoclonal (OKT3) antilymphocyte preparations as a first line treatment for acute rejection of the renal transplant, the majority reserve these preparations for steroid resistant or recurrent rejection episodes. Selection of these agents as first line therapy is more acceptable in liver, heart or lung transplantation, when the cost of graft loss is usually patient death. The decision to cease steroid bolus therapy and commence an antilymphocyte preparation should thus be based on a lack of clinical improvement, evidence of deterioration of blood flow to the transplant, or biopsy evidence of progression of the rejection episode. Both ATG and OKT3 can be given for between ten and 14 days. OKT3 is specifically targeted at T lymphocytes and its effectiveness can be assessed by deletion of T cells from the peripheral blood. Clinical evidence of the effectiveness of OKT3 is variable, with a fall in the serum creatinine often not seen until three days after commencing treatment. A further episode of cellular rejection is frequently encountered within a week of cessation of the course of OKT3, but is usually sensitive to bolus injections of methylprednisolone. The two major limitations to its use are its tendency to precipitate pulmonary oedema if the patient is overloaded with fluid, and the relatively high incidence of

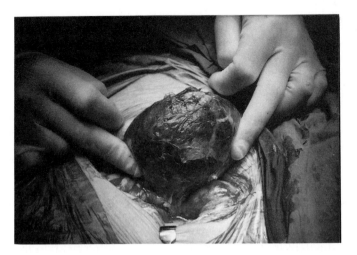

Figure 10.5 Surgical removal of an acutely rejected and subsequently infarcted renal transplant 16 days after transplantation.

cytomegalovirus infection in the subsequent four weeks.

During the end of a course of antirejection treatment, a review of the maintenance immunosuppression is needed. If cyclosporin levels were low, the dose should be increased, and provided there is no evidence of myelotoxicity, the azathioprine dose can be increased to a maximum daily dosage of 2.5 mg/kg. Maintenance prednisolone dosage may be increased to 15 or 20 mg/day.

Early transplant nephrectomy

Renal transplant loss resulting from irreversible acute rejection is usually associated with generalised symptoms of malaise and fever, while the local findings may include discomfort in the region of a thrombosed, inflamed, oedematous and often infarcted graft (see Figure 10.5). Percutaneous biopsy will demonstrate irreversible cortical necrosis and Doppler ultrasound or isotope scanning will confirm absence of cortical blood flow. In these circumstances, early graft nephrectomy is indicated to permit rapid resolution of the symptoms. Under general anaesthesia, the transplant wound is re-opened and an extracapsular mobilisation of the recently transplanted kidney performed. The ureter can usually be avulsed from the bladder, without opening the cystotomy, and the resulting bladder wall defect oversewn with interrupted absorbable sutures. The transplant renal artery and vein should be ligated flush with the external iliac vessels, leaving the vena caval and aortic patches intact.

Acute renal failure

This term is used to describe a diverse group of clinical states associated with acute and severe decline of renal function and is often, but not always, associated with oliguria and, rarely, with anuria. Acute transplant renal failure is usually reversible if the cause is recognised and treated. The immunosuppressed renal transplant patient is commonly confronted with conditions causing acute renal failure, which are grouped into prerenal, renal and postrenal causes.

Prerenal acute renal failure

Reduced perfusion of the renal transplant can result from left ventricular failure, reduced intravascular volume associated with sodium and water depletion or blood loss, acute renal arterial occlusion as discussed later in this chapter, or arterial stenosis as discussed in Chapter 11. Sodium depletion can follow severe diarrhoea or vomiting, overuse of diuretics, uncontrolled diabetes, or poor renal tubular function in the recovery phase of acute tubular necrosis. The pulse rate is usually elevated and the blood pressure low, particularly if taken in the standing position. The increased back resorption of urea causes its concentration in the blood to increase disproportionately to that of creatinine. A low urinary sodium concentration of less than 10 mmol/l and a high urinary osmolality are both pointers to prerenal uraemia. However, diuretics may confound these measures by increasing the urinary sodium and creating a dilute urine. By definition, acute uraemia due to prerenal causes should be promptly reversible when a normal intravascular volume, blood pressure and cardiac output can be re-established. However, if the hypoperfusion of the kidney is either severe or prolonged, the ischaemic injury will cause acute tubular necrosis.

Acute tubular necrosis

Damage to the renal tubules may follow prolonged ischaemia or exposure to nephrotoxic substances. The commonest presentations are either rapid onset of oliguria or a rapid and unexplained increase in the serum creatinine level. Review of the patient's charts may then demonstrate a period of hypotension, perhaps one or two days previously, that may have been associated with an acute episode of left ventricular failure, bacteraemic shock or blood loss. The degree of damage to the renal tubule can be exacerbated by cyclosporin and nephrotoxic antimicrobial agents, dehydration with salt depletion or intravenous contrast media. The diagnosis of ischaemic injury to the tubule can be confirmed by isotope scanning and renal transplant biopsy. The principles of management are to correct blood and fluid deficits, treat sepsis and then maintain normal hydration, avoid nephrotoxic medication, and commence dialysis if necessary. With a simple injury to the renal tubules, regeneration of tubular epithelial cells is usually complete within a matter of days to a few weeks, with urinary concentrating ability slowest to return. More severe tubular injury is associated with disruption of the basement membrane, with subsequent fibrosis and distortion of the renal tubules occurring during the recovery phase and resulting in permanent reduction in renal function.

Postrenal acute renal failure

Obstruction to the urinary collecting system at any point between the calyces and the urethral meatus can cause postrenal acute renal failure. It is essential in all cases of unexplained acute transplant renal failure to rule out obstruction, in which case ultrasound examination will demonstrate a dilated collecting system proximal to an obstruction. Antegrade pyelography may better demonstrate the precise anatomy and cause which may be any one of the acute conditions described later in this chapter under Urinary Tract Complications.

Arterial complications

Early arterial problems are most uncommon, but when they do occur, are generally catastrophic for both the renal transplant and the recipient. Technical errors at the time of surgery, such as poor haemostasis at the point of vascular anastomosis or an intimal flap, are usually apparent at the time of reperfusion of the renal transplant. Rapid correction of the problem should save the renal transplant. An intimal flap may, however, remain unnoticed and result in a renal transplant with primary non-function. Disruption of the suture line can occur in the first weeks after transplantation, especially if local sepsis around the anastomosis weakens the arterial wall. Sudden hypotension, anuria and development of an uncomfortable mass in the region of the renal transplant is indicative of local haemorrhage. Treatment involves urgent surgical exploration and usually graft nephrectomy.

Primary renal artery thrombosis has been described as more likely to occur in the presence of multiple renal arteries, particularly in paediatric size kidneys. Hypoperfusion is a precipitating factor, as may be kinking of a long right sided renal artery. Graft loss is almost inevitable. An embolus into the renal transplant artery will produce painless loss of graft function and is a diagnosis to be considered in patients with a recent myocardial infarction or atrial fibrillation.

Transplant renal vein thrombosis

Acute primary transplant renal vein thrombosis is a controversial diagnosis. Whereas some transplant centres claim that it is virtually non-existent, others report an incidence of between 5 and 10% and implicate cyclosporin associated thrombogenicity as a cause. In our experience, primary thrombosis has occurred in the renal vein of five left sided adult donor kidneys transplanted into adults, and in each case has happened towards the end of the first week after transplantation in functioning renal transplants. Three of the patients also had thrombosis of their arteriovenous fistulas in the days preceding the renal vein thrombosis. Presentation involves painful enlargement of the renal transplant, haematuria proceeding to anuria and followed by graft rupture and hypotension, all within a two to three hour period. All cases were taken immediately to the operating theatre where a patent renal artery and thrombosis of the renal vein were encountered. The operative findings, demonstrated schematically in Figure 10.6, have included a cortical rupture along the convex border of an enlarged, oedematous renal transplant which is surrounded by fresh blood and clot. After thrombectomy of the main renal vein, the renal transplant was able to be recovered on three occasions. Bleeding from the cortical rupture ceased to be a problem after re-establishing normal venous return, and renal function returned after a variable period of ATN.

As a result of these experiences, greater care has been taken to avoid possible kinking by shortening the left renal transplant vein at the time of surgery. Additionally, patients using erythropoietin on dialysis and with haemoglobin concentration of greater than 100 gm/l are given prophylactic subcutaneous heparin 10,000 units per day, while those with acute thrombosis of a fistula soon after transplantation are given 20,000 units of subcutaneous heparin per day. Since adopting this policy, transplant renal vein thrombosis has not been encountered in more than 200 consecutive transplants.

Chronic thrombosis of the transplant renal vein is a separate entity and may be encountered in a graft nephrectomy specimen where the indication for nephrectomy was severe chronic vascular rejection. It usually causes incomplete obstruction of the transplant renal vein.

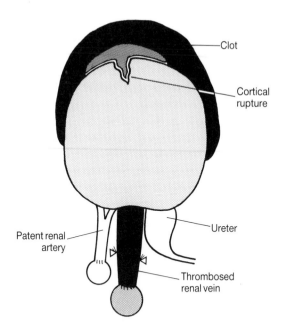

Figure 10.6   Schematic cross sectional representation of operative findings of primary venous thrombosis of a left sided donor kidney transplanted into the right iliac fossa.

Urinary tract complications

Urinary tract complications occur more frequently than vascular complications in the early period after transplantation, with reported incidences varying between 2 and 10% in large series of renal transplant procedures. They are invariably the result of technical error, either at the time of donor kidney retrieval or transplantation, and require additional surgery to correct. Although urological problems are relatively common and potentially serious because of their effect on a solitary functioning kidney in an immunosuppressed patient, they are rarely a cause of graft loss or mortality.

The use of ultrasound examination in the investigation of impaired transplant renal function to distinguish surgical from immunological problems in the early period after transplantation, reduces the morbidity of urological problems by permitting quick recognition and surgical correction.

Extravasation of urine

Avascular necrosis of any portion of the transplanted urinary collecting system can occur after the third day of transplantation and until as late as six months after transplantation. With a non-functioning renal transplant, presentation will be delayed until urine is produced. The site and frequency of extravasation of urine is schematically demonstrated in Figure 10.7. The distal end of the ureter is the most common site since the arterial blood supply comes from the renal artery alone. Damage to the anastomotic chain

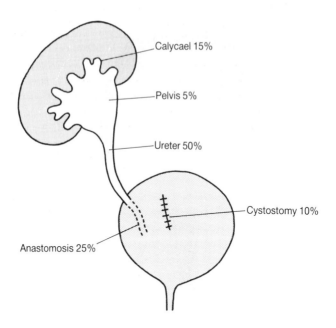

Calycael 15%

Pelvis 5%

Ureter 50%

Cystostomy 10%

Anastomosis 25%

Figure 10.7   Sites of extravasation of urine from the transplant urinary collecting system and their percentage incidence.

of vessels and the support tissue surrounding the ureter at donor retrieval, or use of excessively long ureter at the time of transplantation, will increase the risk of necrosis of the distal end of the ureter. Renal pelvicalyceal ischaemic necrosis may occur if accessory arteries or branches are either ligated inadvertently or anastomosed ineffectively to the recipient arterial system. There does not appear to be an association between rejection episodes and necrosis of the urinary tract. Leakage of urine from the cystostomy can occur any time in the first month after transplantation. Previous bladder pathology, bladder neck obstruction, and the use of high doses of steroids may all contribute to leakage from the ureteric anastomosis or the bladder cystostomy.

Presentation of extravasation of urine is usually dramatic, though this depends on the volume of urine produced and the proportion that leaks from the collecting system. Straw coloured fluid emanating from an external wound drain is common in the first few days after transplantation, and while it is often serous or lymphatic, it may on occasion be urine. Biochemical analysis of the drainage fluid will provide the diagnosis, since urine has a high level of urea and creatinine and usually no glucose. A rapid, but less reliable indication is use of a urine dip stick test to demonstrate the presence of glucose which usually excludes a urine leak. The more common presentation is the onset of oliguria over a few hours, in association with discomfort in the region of the renal transplant. Ultrasound examination will demonstrate a fluid collection in the extraperitoneal space between the bladder and transplanted kidney. While a lymphocele is a differential diagnosis, it is unlikely if the onset of symptoms is rapid. Biochemical analysis of a fine needle aspiration sample of the fluid will produce the definitive diagnosis. Antegrade or retrograde pyelography may assist in localising the site of the extravasation.

The choice of surgical approaches for a necrotic end of ureter depends on the remaining viable length of ureter. Either re-implantation of the ureter onto the dome of the bladder or anastomosis of the adjacent distal ipsilateral native ureter to the pelvis of the transplant ureter are technically satisfactory procedures. In both instances, the use of an internal ureteric stent is advised, avoiding an external draining nephrostomy because of the high risk it carries for bacterial or fungal contamination of the urine collecting system. Vesico-ureteric anastomotic leaks are usually the result of excessive tension on a short ureter. Conversion to a extravesicle ureteric anastomosis is the easiest solution if the original anastomosis was performed in an intravesical position. Otherwise, an end-to-end anastomosis between the pelvis of the transplant renal and the ipsilateral native ureter may be necessary

Extravasation of urine from a calyx is the most devastating urologic complication. In a series of 718 renal transplant patients reported from Brigham and Women's Hospital in Boston, USA, eight of ten patients lost their renal transplant because of this condition and three died of sepsis. It is usually associated with a wedge shaped segment of infarcted renal transplant, and urinary leakage is frequent when more than 10% of the renal mass is involved. Conservative treatment of a calyceal cutaneous fistula has been reported to be successful, as has partial nephrectomy with repair of the calyx and internal stenting of the ureter in our own experience. Without doubt, the best form of management is to prevent renal infarction by ensuring adequate vascularisation of all accessory renal arteries at the time of transplantation.

Early ureteric obstruction

Obstruction of the ureter in the early period after transplantation is uncommon but can be associated with kinking when a redundant length of ureter has been used, or from external compression by a transplant lymphocele. Since most lymphoceles are asymptomatic, before embarking on definitive treatment objective evidence should be obtained before attributing impaired transplant renal function to ureteric obstruction. Intravenous pyelography is often disappointing. Antegrade pyelography or ultrasound guided aspiration of the putative cause of the obstruction may be more helpful in providing objective evidence (see Lymphocele below). Ischaemic stenosis of the ureter usually presents several months after transplantation and is discussed in Chapter 11.

Urinary tract haemorrhage

Haematuria and clot formation are common in the first week after transplantation, are mainly of nuisance value, and are an uncommon cause of impaired renal function. Occasionally, however, large bladder clots require cystoscopy or formal exploration of the bladder. A bleeding vessel may be found at the end of the ureter, or in the bladder wall on the margins of the cystostomy. Bleeding and ureteric obstruction with clot is a risk following percutaneous renal transplant biopsy and is discussed in Chapter 8.

Other causes of obstructive uropathy

Benign prostatic hypertrophy is common in the elderly male recipient, and in patients who have been anuric prior to transplantation. Symptoms may not become apparent until after transplantation and removal of the urinary catheter. After a failed trial of voiding, a urinary catheter should be re-inserted and transurethral resection performed in the fourth week after transplantation. A suprapubic urinary catheter may be necessary if a catheter cannot be passed per urethra. Urethral strictures present similar problems, and if possible, repeated regular urethral dilatations can be attempted rather than a urethroplasty. Diabetics and patients with developmental or traumatic lumbosacral spine pathology may have impaired transplant renal function as a result of inability of the urinary bladder to contract. Either a permanent indwelling catheter or clean intermittent catheterisation is preferable to the formation of an ileal conduit.

Lymphatic complications

Multiple thin walled lymphatic trunks are situated in the retroperitoneal plane anteriorly and on both sides of the external iliac artery and vein. Between 200 and 400 ml per day of clear fluid, with urea and electrolyte concentration similar to that of serum, pass through these trunks to the para-aortic lymph channels and eventually to the venous system. Injury by division or ligation of lymphatic trunks during mobilisation of the iliac vessels in preparation for anastomoses may therefore result in leg swelling, lymphocele formation or a lymphocutaneous fistula. The incidence of these lymphatic complications, nearly always presenting in the first months after transplantation, varies between 1 and 15% depending on whether asymptomatic lymphoceles diagnosed by routine ultrasonography are included in the series. Divided lymphatics draining the renal transplant are another theoretical source of lymph, but investigation of lymphocutaneous fistulae provide evidence that most symptomatic lymphatic problems originate from the recipients' lymphatic trunks (Figure 10.8).

The swollen leg

Transient swelling of the leg ipsilateral to the renal transplant is a common finding in the first week after renal transplantation. On examination, the leg is non-tender and has soft pitting oedema which may also be present to a lesser extent on the contralateral leg. Doppler scanning of the leg veins usually fails to demonstrate a deep vein thrombosis, an uncommon complication in the first week after transplantation. Fluid overload, cardiac failure or hypoproteinaemia may exacerbate the clinical findings. Unilateral oedema usually resolves spontaneously over a matter of days or weeks, and is probably the result of temporary lymphatic obstruction produced

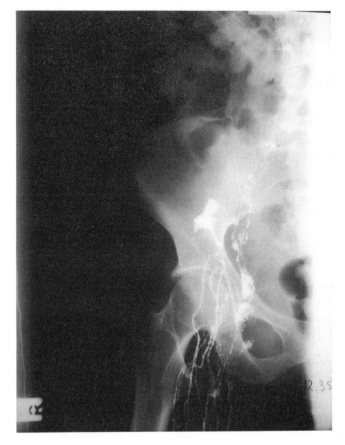

Figure 10.8 Right lower limb lymphangiogram demonstrating leakage of contrast (lipiodol) from a lymphatic trunk in the region of external iliac lymph nodes.

by ligation of lymphatics during transplant surgery. Ipsilateral leg swelling can also indicate the presence of a lymphocele, and can sometimes be associated with severe acute cellular rejection of the renal transplant.

Lymphocele

Lymphatic collections, which are almost invariably situated anterior to the external iliac vessels and inferomedial to the renal transplant, may be unilocular or multilocular. They are usually asymptomatic and can be detected by ultrasound examination. The most common symptom is a vague discomfort in the suprapubic region which is often associated with urinary frequency as a result of bladder compression. There may be a preceding history of excessive volumes of lymph drainage from around the renal transplant in the first week of transplantation or an ipsilateral swollen leg. Probably less than 10% of lymphoceles actually impair transplant renal function as a result of external compression of the transplant ureter. Those that do cause obstruction usually have a volume of greater than 300 ml.

The natural history of the lymphocele is variable, with some resolving spontaneously but the majority remaining asymptomatic. Nocturia as a result of compromised bladder volume is perhaps the commonest indication for intervention (Figure 10.9 a and b). Evidence to support the presence of a lymphocele as the cause of impaired transplant renal function can come from ultrasound examination demonstrating a dilated pelvicalyceal system, isotope scanning demonstrating hold up of tracer in the proximal ureter, intravenous pyelography demonstrating external compression of the ureter, or a diagnostic percutaneous aspiration. Aspiration of the lymphocele on a single occasion may be of diagnostic value and is sometimes curative, with or without instillation of a sclerosant such as tetracycline. Repeated percutaneous aspiration of the lymphocele, however, increases the risk of bacterial contamination. This may preclude subsequent definitive management which involves internal marsupialisation of the lymphocele cavity through a midline abdominal incision. A large window of pelvic peritoneum is excised and a flap of greater omentum placed inside the lymphocele cavity to promote peritoneal absorption of the lymph. Laparoscopic surgery to drain a transplant lymphocele has been described and may be appropriate for unilocular lymphoceles anteromedial to the renal transplant.

Lymphocutaneous fistula

Lymph leakage through the transplant wound is an uncommon but difficult problem to resolve. It is not usually associated with impaired transplant renal function although the chronic daily loss of between 200 and 400 ml of lymph is of nutritional importance. The fistula is usually associated with an ipsilateral swollen leg which is exacerbated by subsequent hypoproteinaemia. Commonly, bacterial or candida infection is present in the fistula wound. A fistula present for more than one month is an indication for lymphangiography to localise the site of lymph leak (see Figure 10.8). Diluted methylene blue can then be injected through the lymphangiogram cannula and the wound explored in the operating theatre. The offending lymphatic trunk is ligated and a peritoneal window created. A drain tube is then placed in the region of the lymphatic surgery, passed into the peritoneal cavity and out through the abdominal wall. The groin wound is then closed primarily and prolonged appropriate antibiotic cover provided.

Drug nephrotoxicity

A host of diverse chemicals have been implicated as nephrotoxins, many of which, for example antibiotics, analgesics and radiological contrast material, have accepted medical uses. When administered under conditions of dehydration and reduced renal perfusion, their nephrotoxic effect may be accentuated. A list of drugs commonly used in conjunction with cyclosporin which may be nephrotoxic appears in Appendix 2. The mechanism of nephrotoxicity can be either direct or indirect by increasing circulating blood levels of cyclosporin.

Cyclosporin nephrotoxicity

The nephrotoxic effects of cyclosporin medication are the result of both functional and morphologic changes, with the most common being the

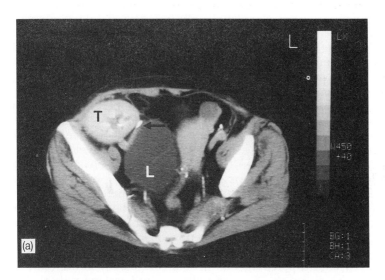

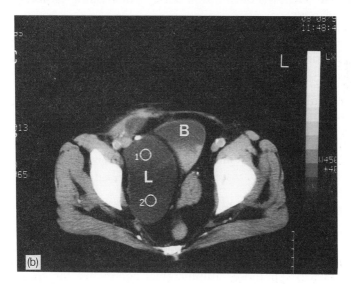

Figure 10.9   (a) Pelvic CT scan (with vascular contrast) in a 40 year old female with urinary frequency and nocturia two months after transplantation, demonstrating a renal transplant (T) in the right iliac fossa and a lymphocele (L) situated inferomedially displacing, but not obstructing, the transplant ureter (arrowed). (b) The urinary bladder (B) is displaced to the left by the multilocular lymphocele (L).

vasoconstrictive effect on afferent glomerular arterioles. The resultant prerenal functional effect of increased vascular resistance is depression of the glomerular filtration rate and renal blood flow. Reduction of flow in the tubular system enhances diffusion of urea out of the tubules and into the blood stream, producing a rise in the serum urea level. Other changes include renal wasting of magnesium and bicarbonate and increased re-absorption of potassium and uric acid. Preservation of serum sodium levels and maintenance of urinary concentrating function, however, suggest that cyclosporin is not associated with generalised tubular damage. This functional effect of cyclosporin is related to the circulating blood level and produces stable impairment of renal function which is initially noticeable towards the end of the first week of transplantation, and is reversible if cyclosporin is discontinued. It is an acceptable impairment when viewed in the perspective of improved renal transplant survival as a result of cyclosporin. If cyclosporin is stopped within the first year after transplantation, the functional changes are reversed and serum creatinine level falls by approximately 25 to 50 μmol/l to establish a new stable level, provided rejection does not intervene as a result of cessation of cyclosporin.

There is a further and less frequent early nephrotoxic manifestation of cyclosporin in which creatinine levels are not stable but rise progressively over days or weeks, and which must be differentiated from rejection. This pattern is usually associated with high blood levels of cyclosporin and responds to dose reduction. A comparison of parameters that may be used to differentiate cyclosporin nephrotoxicity from graft rejection are included in Table 10.2. Acute rejection and cyclosporin nephrotoxicity can co-exist, and a percutaneous needle core biopsy is therefore of value to exclude the diagnosis of rejection before undertaking radical changes in cyclosporin dosage.

Although there are no diagnostic morphologic changes associated with cyclosporin nephrotoxicity, isometric vacuolation and cytoplasmic swelling with proximal tubule cells and tubular microcalcification have been described. Individual smooth muscle cells and endothelial cells of arterioles and small arteries may develop vacuoles, hyalinosis or become necrotic. On reduction of cyclosporin dose or cessation of cyclosporin treatment, renal transplant function will improve as the prerenal functional element of nephrotoxicity is reduced. Mild vascular and tubular morphologic changes will also improve, but occluded arterioles and interstitial fibrosis will not resolve. These irreversible changes may be the basis of chronic cyclosporin nephrotoxicity.

Other nephrotoxic drugs

Renal transplant recipients are often victims of polypharmacy and are prescribed at least one or more direct or indirect (through alteration of cyclosporin metabolism) nephrotoxic drugs in addition to cyclosporin (see Appendix 2). Of these, antimicrobial agents are the most common nephrotoxic drugs, including aminoglycosides, trimethoprim, vancomycin, erythromycin, acyclovir and amphotericin B. Commonly used non-steroidal anti-inflammatory drugs are directly nephrotoxic and should be avoided. Careful monitoring of cyclosporin levels is required in conjunction with diltiazem, ketoconazole and, to a lesser extent, methylprednisolone (see Chapter 7). The antihypertensive agents captopril and enalapril, both angiotensin converting enzyme inhibitor drugs, can be associated with impaired renal transplant function similar to that seen with native renal artery stenosis, and occurring more commonly in renal transplants with already marginal renal function.

Infection

Severe disseminated infection, of any form, is common in immuno-suppressed transplant recipients and has a non-specific prerenal effect on transplant function. Furthermore, antimicrobial agents, particularly those for fungal and viral infections, are nephrotoxic. These issues are discussed in detail in Chapter 13, but two infections require special mention.

Urinary tract infection

Bacterial infections of the bladder are the most common form of infection in the first three months after transplantation, occurring in up to 40% of patients. The adverse effect on renal transplant function is usually indirect and related to the nephrotoxic effect of drugs necessary for treatment. The high risk of urinary tract infection (UTI) is related to pre-existing urinary tract pathology, catheterisation at time of transplantation and

Table 10.2 A comparison of cyclosporin nephrotoxity and acute rejection in cyclosporin treated patients

	Nephrotoxicity	Rejection
History	Poor initial transplant function Other nephrotoxic drugs > 2 weeks after transplantation	< 4 weeks after transplantation
Examination	Afebrile Unchanged transplant size Minimal weight gain Stable urine volume	Low grade fever Enlarged transplant Weight gain Progressive oliguria
Laboratory	Disproportionate urea increase Creatinine rise usually < 50% of baseline Increased K+, uric acid Decreased bicarbonate, Mg++ Elevated cyclosporin levels	Progressive creatinine rise usually > 50% of baseline Low cyclosporin levels
Biopsy	Possible focal infiltrate Intimal thickening Tubular vacuolation	Mononuclear cell infiltrate Interstitial oedema Intimal inflammation Tubular atrophy
Ultrasound	Unchanged cross sectional area	Increased cross sectional area Decreased diastolic blood flow
Therapy	Responds to lower cyclosporin dose	Responds to antirejection treatment

immunosuppression. UTI can be particularly frequent in patients with pre-existing native ureteric reflux. Recurrent episodes of pyelonephritis of the native kidneys are an indication for nephro-ureterectomy. In such instances, it is important that the native ureters be removed flush with the bladder wall.

Most UTI in the renal transplant involves the bladder only, producing symptoms of urinary frequency and suprapubic discomfort. However, despite the surgeon's best intentions, reflux of infected urine into the renal transplant can occur, with transplant pyelonephritis producing high temperatures and an enlarged renal transplant. Biopsy of the renal transplant at this stage will demonstrate interstitial polymorph infiltrate and varying degrees of tubular necrosis (Figure 10.10). UTIs are also the most frequent source of bacteraemia in the first months after transplantation.

Cytomegalovirus

Severe cytomegalovirus infection in the transplant recipient is the result of either primary infection or re-infection. The latent virus is usually transmitted in the renal tubules of the donor kidney. Infection usually occurs in the second or third month after transplantation and is more frequent in patients receiving antilymphocyte preparations. The appearance of cytomegalovirus in the urine or in transplant needle core biopsies is of uncertain significance, but is often seen without evidence of viraemia or impaired renal function. There is no objective evidence to suggest that cytomegalovirus increases the risk of rejection, although a reduction in immunosuppressive therapy necessary to manage leukopaenia may subsequently be associated with rejection.

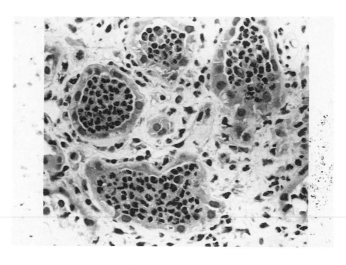

Figure 10.10 Renal cortex biopsy four weeks after transplantation demonstrating renal tubules distended with polymorphs and consistent with acute pyelonephritis (H&E,x128).

Recurrent primary renal disease

Recurrent disease after renal transplantation is usually slow in onset and steadily progressive, but can be encountered as a cause of early transplant dysfunction. Focal segmental glomerulosclerosis may recur within weeks of transplantation, particularly in young patients who have had a previous history of nephrotic syndrome leading rapidly to endstage renal failure and evidence of mesangial proliferation in their native kidney biopsy. The clinical course is marked by heavy proteinuria which is usually progressive, resistant to treatment, and may be so disabling as to require graft nephrectomy. Mesangiocapillary glomerulonephritis type II (dense deposit disease) has a morphological recurrence rate approaching 100% and is usually evident within the first year and as early as three weeks after transplantation. Clinical manifestations of dense deposit disease are slight, however, with graft loss restricted to between 10 and 15% of patients. Membranous glomerulonephritis, IgA nephropathy, antiglomerular basement membrane nephritis and Henoch-Schönlein purpura may each be associated with early renal transplant dysfunction.

Investigation of proteinuria in the transplant patient should include biopsy which is examined by light microscopy, immunofluorescent staining for immunoglobulins and complement components, and electron microscopy. Systemic conditions with renal involvement can also be associated with early renal transplant dysfunction and include haemolytic uraemic syndrome, scleroderma, oxalosis and sickle cell anaemia.

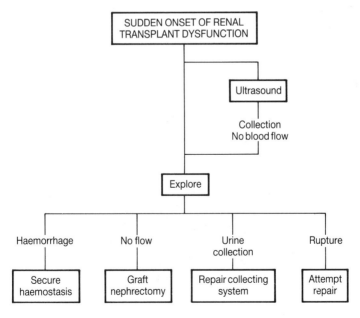

Figure 10.11 Plan for management of sudden onset of renal transplant dysfunction in the early period after renal transplantation.

Management strategies

Both a carefully taken history and a thorough clinical examination are important prerequisites in the management of early transplant dysfunction. There is no place for telephone or corridor consultations. Inappropriate management of rejection compromises patient safety and procrastination may lead to partial or complete and irreversible loss of graft function.

Sudden onset of transplant dysfunction

Development of anuria, pain and graft enlargement over a period of minutes to hours within the first week or two after transplantation are an indication for urgent exploration of the renal transplant. In this situation, a direct line of communication between nursing staff and clinicians responsible for management decisions is of utmost importance to avoid delays that may result in loss of the patient's life, or the renal transplant. An ultrasound examination may be helpful but frequently serves only to postpone the

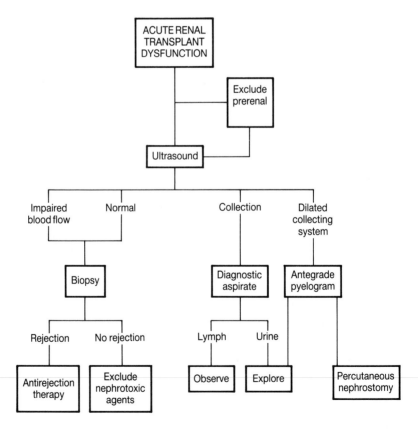

Figure 10.12 Plan for management of acute onset of renal transplant dysfunction in the early period after transplantation.

almost inevitable exploration of the renal transplant. A suggested management pathway is outlined in Figure 10.11.

Acute onset of transplant dysfunction

Less spectacular presentation of transplant dysfunction, with serum creatinine rising from between 20 to 100 μmol/l per day, is due to rejection until proven otherwise. The clinical diagnosis of rejection, however, has been made difficult by the absence of specific symptoms and signs. Ultrasound examination and biopsy are the critical steps in differentiating immunological from other causes of acute transplant dysfunction. Both should be performed on the day in which rejection is suspected, with the major value of the ultrasound examination being the exclusion of mechanical factors. A suggested management pathway is outlined in Figure 10.12 and it should always be remembered that more than one factor may be contributing to the transplant dysfunction.

Slowly progressive transplant dysfunction

The management pathway for investigation of an insidious increase in serum creatinine is not dissimilar to that of acute onset of graft dysfunction. Rejection is again the most important diagnosis to exclude, with ultrasonography and percutaneous transplant biopsy being almost obligatory investigations. It is this group of patients in which cyclosporin nephrotoxicity may be diagnosed. Action on the basis of a raised cyclosporin blood level alone may be regretted, particularly if a co-existing diagnosis of rejection is missed or delayed.

Further reading

1. Olsen TS. Pathology of allograft rejection. In: Burdick JF, Racusen LC, Solez K, Williams GM (eds). *Kidney Transplant Rejection* 2nd edn. New York: Marcel Dekker, page 333–357, 1992.
2. Gruber S A, Fryd D S, Chabers B *et al*. The thrombogenicity of cyclosporin (Editorial). *Clin Transplant* 1988; **2**: 99–101.
3. Jones RM, Murie JA, Ting A *et al*. Renal vascular thrombosis of cadaveric renal allografts receiving cyclosporin, azathioprine and prednisolone triple therapy. *Clin Transplant* 1988; **2**, 122–126.
4. Loughlin K R, Tilney N L, Richie J P. Urologic complications in 718 renal transplant patients. *Surgery* 1984; **297**, 297–302.
5. Jaskowski A, Jones RM, Murrie JA, Morris PJ. Urologic complications in 600 consecutive renal transplants. *Br J Surg*, 1987; **74**, 922–925.
6. Bry J, Hull D, Bartus S A, Schweizer R T. Treatment of recurrent lymphoceles following renal transplantation. *Transplantation*, 1990; **49**, 477–480.
7. Perico N, Renuzzi G. Cyclosporin induced renal dysfunction in experimental animals and humans. *Transplant Rev* 1991; **5**, 63–80.

11 Late renal transplant dysfunction and failure

Much of the research in clinical renal transplantation during the 1970s and 1980s was, understandably, directed towards solving the problem of acute rejection. Rapid losses of renal transplants in the first few months after transplantation were seen as the major barrier to successful transplantation. However, with better immunosuppression and tissue typing techniques, the causes of renal transplant loss in the first six months after transplantation in the 1990s have now become equally distributed between rejection, non-immunological factors, and patient death with a functioning graft (Chapter 10, Figure 10.1). The improvement in these results has redirected emphasis to maintenance of long term graft function and the need to reduce loss of grafts, estimated at 3 to 5% of those at risk each year after transplantation. In the first five years after transplantation, graft loss is more likely to result from chronic rejection. Thereafter, death of the patient accounts for the majority of losses between five and 20 years after transplantation. Until solutions are found to the various causes of graft loss, renal transplantation cannot be considered a truly effective long term treatment option for endstage renal disease.

Research into the causes and treatment of chronic graft failure has proved difficult because of the long time interval required for investigation of aetiological factors. Furthermore, statistically valid differences between treatment regimens require large numbers of patients. Since 50% of the grafts lost in most series are due to patient death with a functioning graft, technical failures or non-compliance with immunosuppression, it is difficult to subject the phenomenon of chronic graft dysfunction to controlled trials. Non-immunological causes, however, may be related indirectly to the immunosuppression protocols if they have an impact on the incidence of coronary artery disease, infection or malignancy.

Large multicentre studies have in the past been, and will continue to be, the main vehicle for identification of factors associated with chronic transplant failure. As with the Collaborative Transplant Study for example, this can minimise the centre to centre differences in patient selection and treatment protocol, but has the disadvantage of being able to collect only a relatively small amount of simple data on a reliable basis. For this reason, research into chronic graft failure has been subject to relatively simplistic analyses of treatment protocols that are usually out of date before long term analysis is achieved. Patients followed for ten to 20 years after transplantation have received azathioprine and prednisolone, a treatment regimen associated with a high incidence of early graft loss due to acute rejection. With the introduction of cyclosporin, the pattern of graft loss has changed dramatically. One approach to the problem of examining variables of long term graft survival is demonstrated in Figure 11.1 in which the outcome of currently observed trends in tissue typing are extrapolated to estimate

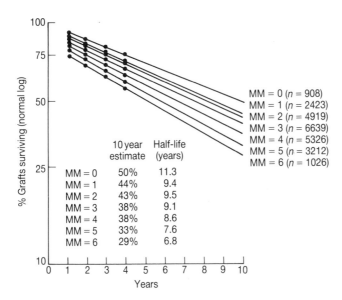

Figure 11.1 Projected ten year cyclosporin treated renal transplant survival according to the number of HLA-A, -B and -DR mismatches (data from the Collaborative Transplant Study, December 1990).

the long term results. The important conclusion from these multicentre data is that time consuming and expensive tissue typing and organ sharing schemes should not be abandoned merely because of what appear to be only small gains in graft survival when analysed one year after transplantation.

In this chapter, the causes of graft loss after the initial months of transplantation are described, with emphasis placed on identification of reversible causes (Table 11.1), and management of the transplant recipient with a failing or failed graft.

Chronic rejection

Chronic rejection is the designated term for slow, progressive and seemingly uncontrollable loss of renal transplant function. It is a poorly understood phenomenon associated with histological evidence of tubular changes consistent with progressive ischaemia. The term presupposes that there is insufficient immunosuppression to control the immunological response of the recipient to the renal transplant. All analyses of transplant patients, whether multicentre or from individual units, demonstrate a steady rate of decline of long term functioning renal transplants, which despite improvements in immunosuppression modalities has not changed in the last two decades. Cyclosporin has served to improve the transplant survival curve by reducing early graft loss and has probably affected the rate of long term graft loss. There is also evidence to suggest that the rate of decline in renal transplant function towards the need for dialysis and retransplantation can be improved by increasing the level of immunosuppression, by providing optimal control of blood pressure and, perhaps, by restriction of dietary protein.

Risk factors

The least controversial risk factor is the degree of histocompatibility locus antigen matching, the effect of which is demonstrated in Figure 11.1. Projected figures demonstrate a 21% difference in graft survival between

Table 11.1   Causes of late renal transplant dysfunction

Prerenal	Transplant renal artery stenosis
	Recipient aorto-iliac occlusive arterial disease
	Patient death
Renal	Chronic rejection
	Cyclosporin nephrotoxicity
	Other nephrotoxic drugs
	Immunosuppression withdrawal
	Transplant pyelonephritis
	Recurrent primary renal disease
	De novo glomerulonephritis
Postrenal	Transplant ureteric stenosis
	Transplant vesico-ureteric reflux
	External compression of ureter by lymphocele
	Neurogenic bladder
	Prostatic hypertrophy
	Urethral stenosis

well matched and poorly matched kidneys ten years after transplantation (50 vs 29%). If nephrons for an individual renal transplant are lost at a steady rate as a result of chronic rejection, it follows that a renal transplant will function longer if more nephrons are present at the time of transplantation. This concept is supported by evidence that kidneys retrieved from younger donors survive longer than those from old donors, irrespective of the age of the recipient (Figure 11.2). The independent influence of age of recipient on graft survival relates to the increased risk of cardiovascular disease and hence death with a functioning graft. There has also been a suggestion that children less than six years of age have a more competent immune system, and are therefore more likely to sustain graft loss for immunologic reasons.

Clinical presentation

The insidious decline of graft function, to the stage of requiring dialysis, usually occurs over several years, but may be more rapid with graft loss in only a few months. A continuous decline in renal transplant function may be identified if assessed by measured glomerular filtration rate (GFR); changes in creatinine level, on the other hand, depend on the size of the patient and their dietary intake and are influenced by intercurrent illness. An increase in serum creatinine will not usually be seen until the GFR is less than 30 ml/min/1.73m^2. Although multiple or severe episodes of treatable acute rejection are more likely to be followed by chronic graft failure, early graft failure is not inevitable. As an expression of the unpredictable nature of 'chronic rejection', long term excellent graft function may be seen after severe early rejection and equally, progressive loss of transplant function may be observed in patients who have never suffered an episode of acute rejection.

The initial clinical presentation of chronic rejection, as assessed by serum creatinine, may be associated with events, such as volume depletion, bacteraemic shock and exposure to nephrotoxic agents, that precipitate acute renal failure. Although the renal transplant may recover from these events, it will usually be at the cost of further loss of functioning nephrons. As transplant function declines, markers of chronic renal failure such as

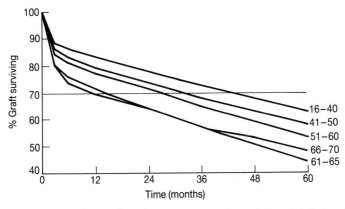

Figure 11.2 Effect of renal transplant donor age on cyclosporin treated first cadaver renal transplant survival (data from the Collaborative Transplant Study, May 1992).

anaemia, acid base imbalance and abnormal bone metabolism appear. On examination, the renal transplant becomes smaller and analysis of urine, may demonstrate increasing proteinuria.

Influence of immunosuppressive agents

The relationship between chronic rejection and choice of immuno-suppressive agents is much debated. Recent data from the Collaborative Transplant Study, however, suggests that the advantage for cyclosporin, in terms of graft survival, is not just in the short term, but also in the long term (Figure 11.3).

The adverse effect of long term exposure of the renal transplant to cyclosporin and its effect on afferent arterioles is poorly understood and not easily differentiated from the arterial injury of chronic rejection. In models of progressive glomerulopathy involving renal obliteration, corti-costeroids exacerbate the glomerular capillary hypertension. Inadequate immunosuppression, however, may be responsible for unremitting but subclinical immune destruction of the renal transplant, while acute rejection can occur at any time after transplantation if immunosuppression is reduced or withdrawn.

Histopathology of chronic rejection

Histopathological changes attributable to acute or to chronic rejection are not well demarcated. Both processes can co-exist and occur at almost any time after transplantation. The most notable features of chronic rejection are occlusive changes to intrarenal arteries, together with tubular atrophy and interstitial fibrosis which are presumed to be the result of vascular insufficiency. On macroscopic examination, the surface of the kidney is pale, and with increasing time after transplantation, the kidney becomes smaller and the distinction between the cortex and medulla less obvious.

Arteries of any size, either in the hilum or intrarenal, can be affected, although the interlobular and arteriolar vessels are the most severely affected by concentric fibroblastic thickening of the intima (Figure 11.4). Although

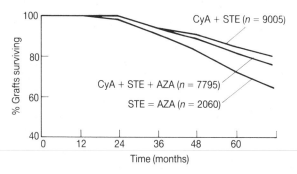

Figure 11.3 Survival rates of first cadaver transplants, after one year of graft function, in patients on different combinations of immunosuppressive drugs. (CyA = cyclosporin, STE = corticosteroids, AZA = azathioprine) (data from the Collaborative Transplant Study, August 1992).

more commonly seen in acute rejection, fibrinoid necrosis of the media may occur, but disruption of the internal elastic lamina and atrophy of the smooth muscle are frequent findings. Deposition of immunoglobulins and complement, demonstrated by immunofluorescent staining, are inconstant but suggestive of an antibody mediated immunological response. The simplified sequence of events which create these changes is thought to be initiated by damage to the vascular endothelium, either by recipient antibody or antigen–antibody complexes and complement. Exposed collagen activates the clotting cascade and a complex inflammatory response within the arterial media follows. The repair process resulting from this inflammatory response leads to gradual obliteration of the lumen.

The glomeruli may become smaller while developing thickening of the glomerular capillary basement membrane and areas of focal hyalinisation. There are many possible causes for these glomerular changes, which are variable in degree but collectively described as transplant glomerulopathy, including immunologic responses, ischaemia and glomerular hyperfiltration injury.

Progressive atrophy of renal tubular cells together with luminal obliteration are prominent features of chronic rejection and reflect progressive ischaemia of the graft (see Figure 11.4). These changes are associated with fibrosis of the interstitium, which is a non-specific response to ischaemia and can occur in the absence of a mononuclear cell infiltrate. If these cells are present, however, and particularly if they can be seen to be infiltrating renal tubules, they may be indicative of inadequate maintenance immunosuppression. The mononuclear cells seen in chronic rejecting grafts are a mixture of T and non-T lymphocytes, while the cells seen during acute rejection are predominantly T cells.

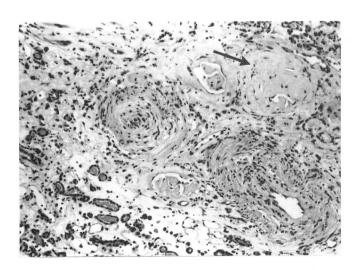

Figure 11.4 Chronic vascular rejection affecting two interlobular arteries, one of which is occluded by intimal fibrous thickening. Also demonstrated in this cortical biopsy is interstitial fibrosis, tubular atrophy and a sclerosed glomerulus (arrowed) (H&E,x64).

Diagnosis

The diagnosis of chronic rejection is based on histopathological findings in a needle core biopsy together with exclusion of other possible causes of late transplant dysfunction. Late ureteral stenosis and renal artery stenosis are important but uncommon causes of chronic failure mandating ultrasound examination of the renal transplant. Angiography may be necessary to make the latter diagnosis. Recurrent primary renal disease may mimic chronic rejection but can usually be distinguished on biopsy. There are no reliable features by which to distinguish chronic cyclosporin nephrotoxicity from chronic rejection, and a trial period of reduction of cyclosporin dose or conversion to azathioprine and prednisolone therapy may be necessary to distinguish this condition.

Management

There is no simple way to avert graft loss from chronic rejection, although three treatment strategies are worthy of consideration. Firstly, the clinical presentation and histological findings of chronic rejection may follow either continuous subclinical or recurrent frank episodes of acute rejection. If so, increasing maintenance immunosuppression may help to maximise long term function. In patients treated with triple therapy, an increase in the dose of cyclosporin can be judged by measurement of blood levels, but may further complicate assessment of graft function. An alternative is to increase the azathioprine dosage of 2.5 mg/kg/day, or prednisolone to between 15 and 20 mg/day. In the absence of a biopsy demonstrating an acute interstitial cellular infiltrate with other evidence of acute rejection, neither bolus steroid therapy nor antilymphocyte preparations are beneficial.

The second treatment strategy involves intervening in the repair mechanisms responsible for intimal thickening and arterial occlusion. Many transplant clinicians are able to provide at least anecdotal evidence of the benefit of dipyridamole, aspirin or warfarin. A role may in the future be found for specific inhibitors of the mediators of inflammatory responses.

The third treatment strategy is directed at reduction of the effect of the non-immunologic factors associated with decline in renal function. Progressive narrowing of afferent arterioles is usually accompanied by worsening hypertension. Maximum control may therefore be of benefit even though the choice of antihypertensive agents is often difficult. Whereas angiotensin converting enzyme (ACE) inhibitors have, in theory, the advantage of reducing the filtration stress on remaining glomeruli, the result in the transplanted kidney with arteriolar narrowing is to further reduce the GFR. An alternative is to use combinations of less effective antihypertensive agents such as calcium channel blockers, beta blockers, hydralazine, minoxidil and clonidine. If these prove ineffective, ACE inhibitor therapy may be necessary, but should be introduced in small dose increments in association with frequent review. Restriction of dietary protein to reduce the filtration stress on remaining glomeruli may have a role, though this remains unproven.

Chronic cyclosporin nephrotoxicity

The major impact of cyclosporin on renal transplantation has been the approximately 80% reduction of acute rejection as a cause of graft loss in the first six months after transplantation. This benefit far outweighs the acute functional nephrotoxic effect of cyclosporin. Thereafter, withdrawal of cyclosporin is associated with a small but real incidence of graft loss from acute rejection which has influenced most transplant clinicians to maintain patients on cyclosporin for the life of the renal transplant. There is, however, concern over the continuing and indefinite use of a drug known to be nephrotoxic, and there is much debate as to whether or not cyclosporin is associated with progressive and irreversible nephrotoxicity. On clinical and histopathological grounds, it is often difficult to differentiate cyclosporin nephrotoxicity from chronic renal transplant rejection; however, a chronic injury to the native kidneys is seen in cardiac and liver transplant recipients and in patients with auto-immune diseases receiving cyclosporin therapy. On balance, most centres currently favour long term use of cyclosporin, provided appropriately and individually adjusted patient dosage schedules are employed. Nevertheless, a definitive statement on the possibility of progressive cyclosporin associated renal injury cannot be made until completion of long term studies of renal and other solid organ transplant recipients.

Clinical manifestations of chronic nephrotoxicity

There is no doubt that when transplant recipients receiving azathioprine and prednisolone alone are compared with those given cyclosporin, there is an associated decline in GFR in the first six to 12 months. Proponents of the safe long term use of cyclosporin argue that the GFR and serum creatinine remain stable thereafter, providing evidence to support the claim that cyclosporin induced deterioration in renal function is not progressive. Furthermore, they argue that the slow rate of decline of renal transplant function in the presence of cyclosporin is a reflection of chronic immunologic injury, and that this rate of renal transplant attrition is greater in the absence of cyclosporin therapy. Opponents, however, argue that serum creatinine is a poor monitor of renal transplant function and that stabilisation of GFR in the presence of cyclosporin is the result of compensatory changes in the remaining glomeruli. If this is true, long term histological assessment would be expected to demonstrate progressive glomerular sclerosis. It would be reasonable, at present, to conclude that the effect of cyclosporin on long term outcome is not known.

Histopathology of chronic nephrotoxicity

If separate and specific histological changes existed that could differentiate chronic renal transplant rejection from cyclosporin associated renal injury, there would be no argument of whether the latter exists. In renal transplants, both conditions can co-exist, and the renal lesions seen in patients receiving cyclosporin for conditions other than renal transplantation are difficult to differentiate from the independent vascular changes associated with hypertension or diabetes. In general, however, if vascular narrowing or occlusion is predominantly restricted to afferent arterioles with two or less layers of smooth muscle cells, the injury is more likely to be the consequence of cyclosporin therapy. This may be associated with tubular atrophy; interstitial fibrosis which may be diffuse or distributed in stripes; or segmental and global glomerular sclerosis, corresponding to areas of cortex associated with injury to the afferent arteriole (Figure 11.5). An additional feature is replacement of arteriolar smooth muscle cells by hyaline deposits. Chronic rejection, on the other hand, is more likely to be associated with more widespread vascular injury with concentric narrowing of the arterial lumen.

Management

Whilst the differentiation of chronic cyclosporin nephrotoxicity from chronic rejection remains a dilemma, the therapeutic options for improving graft function in a long term renal transplant patient remain limited. Hypertension associated with cyclosporin is dose related and tends not to be progressive in severity. The use of diuretics in these patients may exacerbate the prerenal effect of cyclosporin and the use of peripheral vasodilators is preferable. Withdrawal or substantial reduction of the cyclosporin dose may produce impressive initial improvement in graft function and control of hypertension, but more than 40% of these

Figure 11.5 Renal transplant biopsy 12 months after transplantation demonstrating band-like or 'striped' patchy interstitial fibrosis associated with tubular atrophy (H&E,x16).

patients will develop acute rejection that may not appear until months after alteration to the cyclosporin dose. Careful and prolonged monitoring after major reduction of the cyclosporin dose is therefore indicated.

Patient death with a functioning graft

After the first five years of renal transplantation, graft loss as a result of death of the transplant recipient outnumbers loss from chronic rejection by two to one (see Chapter 14). The numerical importance of death of the renal transplant recipient, the principal cause of which is cardiovascular disease, is such that it is now the major limiting factor to successful long term renal transplantation. The problem can be addressed in several ways, and while cadaver renal transplants remain a limited resource, there is the temptation to limit transplantation to patients less than 45 years of age, since this would reduce graft loss following cardiovascular mortality. However, this would be unacceptable to most clinicians solely on ethical grounds. A compromise solution is to screen potential transplant recipients for significant cardiovascular disease, such that those patients may first have their cardiac disease treated. After transplantation, it is then important to identify and aggressively treat risk factors for cardiovascular mortality such as hyperlipidaemia and diabetes.

The risk of developing malignancy increases with time after transplantation and surveillance by transplant clinicians, particularly for skin tumours and genito-urinary tract tumours is obviously important. The pattern of lymphoproliferative disorders after transplantation has changed with the combination of triple immunosuppressive maintenance therapy and the variety of antilymphocyte preparations. Death from uncontrolled lymphoproliferative disorders is now more likely to occur in the first one to two years after transplantation rather than later. Overall, death from malignancy or infection is of much less importance in the long term transplant recipient than mortality associated with coronary artery disease and cerebrovascular accident. The role of the immunosuppressive agents in the development of atherosclerosis is therefore an important reason for including patients dying with a functioning graft in analyses comparing the efficacy of various immunosuppressive agents and regimens.

Transplant renal artery stenosis

Hypertension is common after successful renal transplantation, occurring in more than half the recipients. Transplant renal artery stenosis is however, an uncommon cause, with the reported incidence varying from one to 20%, reflecting the differing criteria for diagnosis. The importance of making a diagnosis of transplant renal artery stenosis is that it is a reversible cause of late transplant dysfunction and a preventable cause of late graft loss. Surgical or radiological intervention is necessary in between 1 to 3% of renal transplant recipients.

Patterns of presentation

Median time of presentation is between six and 12 months after transplantation. Perhaps a third of patients present with simultaneous onset of graft dysfunction and hypertension, while in the remainder, hypertension precedes graft dysfunction by several months or more. Patients usually require a combination of three or more antihypertensive agents, with the diagnosis often becoming apparent after introduction of an ACE inhibitor which decreases renal plasma flow dramatically.

The finding of a bruit on auscultation over the renal transplant is common after renal transplantation and may indicate either a flow disturbance in the transplant renal artery, an arteriovenous fistula in the kidney, or may emanate from the recipient iliac artery. A bruit is usually indicative of a benign diagnosis, and conversely, is present in less than 50% of patients presenting with renal artery stenosis requiring intervention. The appearance of a bruit as a new finding months to years after transplantation in the presence of poorly controlled hypertension is, however, a significant finding.

Diagnosis and investigation

Many authors believe that all patients with poorly controlled hypertension warrant investigation by angiography. Less invasive investigation is possible by the use of test doses of an ACE inhibitor and duplex ultrasonography. The latter demonstrates exaggerated systolic wave forms together with increased flow velocity across the stenosis. Angiography should be performed with both anteroposterior and oblique views of the transplant renal artery. There are four distinct types of stenosis based on angiographic findings. The first type involves the renal artery anastomosis and is possibly technical in origin. It is most likely to occur after an end-to-end anastomosis of the internal iliac artery to the transplant renal artery, rather than end-to-side anastomosis to the external iliac artery using the Carrel aortic patch technique. Anastomotic stenoses were more common when black silk sutures were used rather than the current monofilament synthetic suture material. The incidence of this type of stenosis has thus been remarkably reduced in recent years.

The second and more frequent type of stenosis is postanastomotic (Figure 11.6). The aetiology is probably multifactorial and includes physical damage to the endothelium, vascular rejection and external compression by the recipient's fibrous reaction to the presence of the renal transplant. The stenosis is usually fibrotic and non-atheromatous, commonly involving much of the length of the main renal artery to the point of secondary branching. Kinking of the transplant renal artery following poor positioning of the transplant is also included in this group and interestingly, post-anastomotic stenoses are twice as common in the longer right transplant renal artery.

The third type of stenosis is associated with chronic rejection and is more common in patients with multiple episodes of acute rejection. Multiple stenoses are seen in the secondary and tertiary branches of the renal artery, often giving them a beaded appearance on angiography (see Figure 12.15).

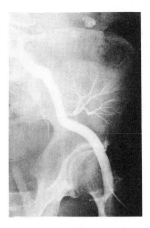

Figure 11.6   Postanastomotic transplant renal artery stenosis in a right sided donor kidney transplanted using an end-to-side anastomosis to the left external iliac artery of the recipient.

The fourth and perhaps less frequent type occurs in the recipient arteries proximal to the transplant renal artery stenosis. The aetiology is atherosclerosis and although not strictly renal artery stenosis, occlusive peripheral vascular disease in the native arteries supplying the renal artery produces the same functional effects (Figure 11.7).

Management

Angiography, although not always able to provide a definitive diagnosis, is necessary before contemplating interventional management of transplant renal artery stenosis. The indications for angiography are listed in Table 11.2. Once the stenosis is identified, its functional significance may need to be demonstrated. A stenosis of less than 50% is probably not significant and other causes for hypertension are a more likely cause of the hypertension.

When the renal artery stenosis is greater than 50%, a renal transplant biopsy can assist decision making. The presence of tubular atrophy, crowding of the glomeruli and cortical shrinking are indicative of prerenal ischaemia. On the other hand, presence of chronic rejection with widespread arterial and arteriolar changes of concentric luminal narrowing and interstitial fibrosis are a contraindication to interventional management of renal artery stenoses, particularly when involving multiple

Table 11.2   Indications for angiography to investigate possible transplant renal artery stenosis

- Worsening hypertension despite increased use of antihypertensives
- Renal transplant bruit and worsening hypertension
- Deterioration in transplant function after introduction of ACE inhibitor therapy
- Duplex ultrasonography or ^{99m}Tc DTPA scan diagnosis of renal artery stenosis

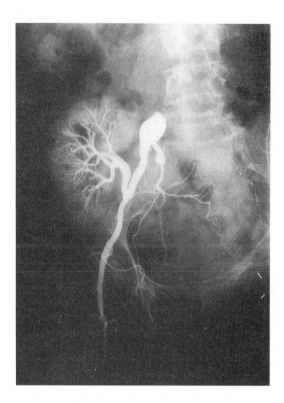

Figure 11.7 Angiogram of a renal transplant five years after transplantation demonstrating occlusion and aneurysm formation of the right common iliac artery. The transplant recipient was hypertensive and had been anuric for two months. After revascularisation, achieved by aortobi-iliac grafting, urine was produced within one hour of completion of the anastomoses and normal renal function returned.

renal artery branches. Additional evidence for the significance of a stenosis can be gained by introducing small doses of an ACE inhibitor. Alternatively, measurement of the systolic gradient across the stenosis can be performed at the time of angiography, with advocates of this technique claiming that successful correction of hypertension is possible when the systolic pressure gradient is greater than 60 mmHg.

The treatment options for interventional management are percutaneous transluminal angioplasty (PTA) or surgery. Both methods are probably of equal effectiveness for blood pressure control but PTA entails a higher rate of initial failure and frequent restenosis because of the fibrotic nature of the arterial lesion. Nevertheless, because it is both technically easier and better tolerated by patients, PTA in centres with appropriate experience is considered to be the treatment of first choice for correction of transplant renal artery stenosis. The best results are achieved with PTA when treating postanastomotic stenosis.

Surgical correction of transplant renal artery stenosis is both tedious and technically challenging and is indicated when PTA is not feasible or has failed. Depending on the position of the transplant renal artery, an intraperitoneal approach rather than an incision through the old transplant wound, is usually more appropriate. If the stenosis is proximal, an end-to-end anastomosis to the internal iliac artery or an end-to-side anastomosis to the external iliac artery may be possible. If not, an interposition saphenous vein graft is preferable to a vein patch. The renal transplant can usually withstand 20 to 30 minutes of warm ischaemia while these anastomoses are being performed. Reported success rates are between 60 and 70%. Atherosclerotic disease in native arteries proximal to the renal transplant is treated on merit by either PTA, endarterectomy or bypass surgery. In the presence of severe and extensive aorto-iliac occlusive disease, it may be necessary to anastomose the end of the transplant renal artery to the side of the dacron bypass graft.

Late transplant urinary tract complications

Whereas urinary tract complications in the first few weeks after transplantation are more common and technical in origin (see Chapter 10), complications occurring thereafter are more insidious in presentation, and their aetiology less clear. With a reported incidence of between 2 and 10%, they are an important reversible cause of late renal transplant dysfunction.

Transplant ureteric stenosis

Progressive stenosis of the transplant ureter can present months to years after transplantation and is manifest by slowly deteriorating renal function. The two principal sites of stenosis are the distal end of the ureter and the pelvi-ureteric junction, and each has a different probable aetiology. Ischaemia to a degree that is insufficient to cause ureteric necrosis is a logical cause of distal transplant ureteric stenosis, occurring more commonly in situations where the blood supply to the transplant ureter is likely to be compromised. It is unusual to demonstrate evidence of rejection in the distal ureter if it is subsequently excised. The alternative explanation is peri-ureteric fibrosis that might occur as a local host reaction to the presence of the transplanted ureter, either in the submucosal tunnel of the urinary bladder or in an extravesical portion. Pelvi-ureteric obstruction usually occurs within the first 12 months after transplantation. Peri-ureteric fibrosis is the probable cause of obstruction at the pelvi-ureteric junction, with the distal ureter being normal. Unless caused by a lymphocele (discussed in Chapter 10), external compression of the transplant ureter is uncommon.

The key to diagnosis of a ureteric stenosis is hydronephrosis demonstrated on ultrasound examination (Figure 11.8). Mild to moderate dilatation of the pelvicalyceal system is frequent after transplantation and is more often than not associated with normal transplant renal function. Sequential ultrasound examinations demonstrating progressive dilatation of the pelvicalyceal system, however, are an indication for further investigation. A physio-

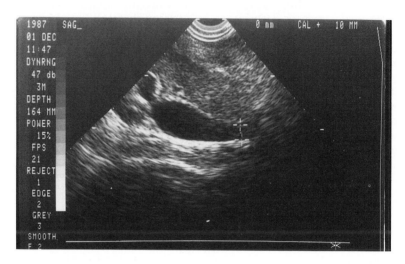

Figure 11.8 Ultrasound image of a renal transplant demonstrating hydronephrotic calyces and renal pelvis.

logically significant dilated pelvicalyceal system will be associated with increased intrarenal pressure. The arterial wave forms demonstrated on duplex ultrasonography will thus be dampened, and when hydronephrosis is severe, will be associated with absent diastolic flow. Relief of the obstruction by percutaneous nephrostomy and rapid reduction in intrarenal pressure is followed by return of more normal arterial wave forms within minutes.

If the serum creatinine is less than 250 μmol/l, intravenous pyelography may be of benefit, but usually fails to demonstrate the site of ureteric stenosis with accuracy. Equally, if transplant renal function permits, a DTPA nuclear scan together with frusemide to accentuate the hydronephrosis, while providing supporting evidence of a significant ureteric stenosis (Figure 11.9), does not provide good anatomical information.

Retrograde ureteropyelography is often unsatisfactory because of difficulty passing a catheter into the distal transplant ureter at cystoscopy. For this reason, antegrade pyelography is the investigation of choice to provide good anatomical data. After administration of prophylactic antibiotics, and using ultrasound localisation, an anterior mid or upper renal transplant calyx is punctured with a 21 gauge needle. The pelvis and ureter are opacified and the stenotic segment identified. Urodynamic studies are not necessary as the findings at antegrade pyelography are usually unequivocal. A guidewire is then placed in the collecting system of the renal transplant and an 8.0 French pig-tail catheter inserted for external drainage to provide temporary relief of the obstruction and correction of uraemia (Figure 11.10). After several days of external drainage, the ureteric stenosis can often be crossed by a guidewire and a double J stent subsequently placed across the stenosis over the guidewire. The external nephrostomy tube is left in place for a further day or so before removal after a period of clamping to test the patency of the internal double J stent. Elective surgical repair of the stenosis can then be performed at a later date.

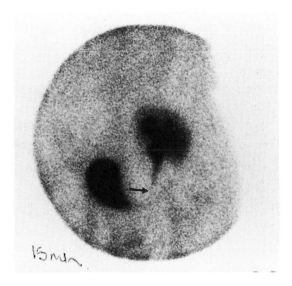

Figure 11.9 DTPA nuclear scan demonstrating hold-up of tracer in the renal pelvis of a left sided renal transplant. In addition, there is displacement and compression of the transplant ureter (arrowed) and compression of the urinary bladder by a lymphocele seven years after transplantation.

The surgical management of ureteric stenosis is dependent on the length of viable remaining transplant ureter. Most transplant surgeons advocate an intraperitoneal approach to the transplant ureter, with identification of the ureter facilitated by the presence of a ureteric stent. Provided sufficient transplant ureter length is available, one option is to re-anastomose the ureter to the dome of the bladder. Nevertheless, this dissection is often difficult because of the presence of reactive fibrous tissue surrounding the transplant ureter. The alternative is to anastomose the distal native ureter, if present, to the transplant renal pelvis or viable proximal transplant ureter (Figure 11.11). The anastomosis is protected by a double J stent which is removed at cystoscopy four to six weeks after surgery.

Whichever surgical option is taken, the procedure is technically difficult to perform and is associated with a reported complication rate of about 25%. Urine leaks and infection are common, particularly if urinary tract infection is present prior to the surgery. Percutaneous nephrostomy drainage tubes should be avoided after surgery. The technical difficulty of surgery and the high complication rate have prompted some transplant units to advocate percutaneous dilatation of the ureteric stenosis at the time of, or soon after, percutaneous drainage of the obstruction. If successful, a pig-tail catheter is left in position for four to six weeks. The restenosis rate, however, is high and the procedure is usually unsatisfactory for obstruction of the pelvi-ureteric junction. For the latter, either pyelostomy or anastomosis of the native ureter to the transplant renal pelvis is necessary. Provided the native kidney on that side is non-functioning and has not been associated with infection, the ureter can be divided and ligated without nephrectomy.

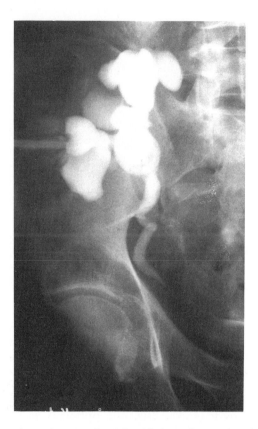

Figure 11.10 Antegrade pyelogram of a right sided renal transplant five years after transplantation and demonstrating pelvicalyceal dilatation and a tapered distal transplant ureteric stenosis.

Transplant vesico-ureteric reflux

The reported incidence varies widely between one and 40%, reflecting both the frequency with which investigation is undertaken and the care with which the vesico-ureteric anastomosis is performed. Vesico-ureteric reflux is frequently asymptomatic and not associated with deteriorating renal transplant function. Recurrent urinary tract infections, particularly transplant pyelonephritis, occurring despite appropriate antibiotics prophylaxis are an indication to perform a micturating cysto-urethrogram (Figure 11.12). If reflux is demonstrated and infection remains uncontrolled, re-implantation of the transplant ureter should be considered. The long term effect of vesico-ureteric reflux on the transplant renal function is uncertain but may be harmful.

Other urinary tract pathology

Functional obstruction of the native urinary tract can follow urethral

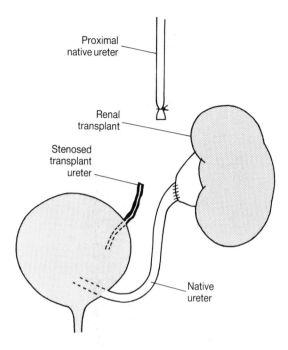

Figure 11.11 Anastomosis of the distal native ureter to the pelvis of the renal transplant for surgical correction of a stenosed transplant ureter. The proximal half of the native ureter has been ligated and the native kidney left *in situ*.

stenosis, prostatic hypertrophy or malignancy, or a neurogenic bladder as a result of spinal cord pathology or diabetic autonomic neuropathy. Urethral stenosis is more common in males and may have pre-existed transplantation or followed prolonged use of a urethral catheter after transplantation. Treatment is by urethral dilatation at regular intervals or urethroplasty. The increased safety of transplantation of the elderly and improved graft survival provided by cyclosporin has made prostatic hypertrophy a more common problem in the male renal transplant population. As yet, the reported numbers of male patients with prostatic malignancy is small and it is not clear yet if there is a significantly increased risk after transplantation.

Renal transplant calculi are unusual, although there is an increased incidence related to the use of ureteric stents, the presence of tertiary hyperparathyroidism and hyperuricosuria. The presenting symptoms for patients with renal transplant calculi are deteriorating renal function due to obstruction and/or haematuria. Because the renal transplant and ureter are denervated, renal colic is not a presenting symptom, but discomfort in the region of the renal transplant together with enlargement of the renal transplant may occur as a result of acute obstruction.

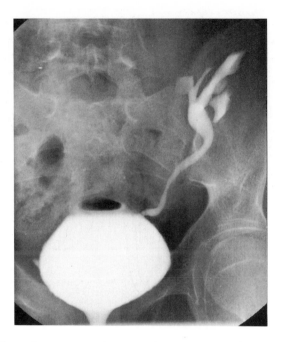

Figure 11.12 Micturating cysto-urethrogram demonstrating reflux of contrast material into the urinary collecting system of a left sided renal transplant seven years after transplantation.

Recurrent primary disease

The reported incidence of graft failure secondary to recurrent disease is less than 2%, although this may be an underestimate because of the difficulty in establishing this diagnosis conclusively. Reasons for the underestimation include an unknown diagnosis of primary renal disease and the difficulty in differentiating co-existing glomerular rejection and glomerulonephritis. Confirmation of the diagnosis of recurrent glomerulonephritis is made on the basis of examination of a transplant biopsy by light microscopy, immunofluorescent staining and electron microscopy (Figure 11.13).

Almost all types of glomerulonephritis have been reported to recur after transplantation and there is considerable variation between the types of glomerulonephritis and the frequency of recurrence, clinical pattern and prognosis (see Tables 2.2 and 11.3). The first clinical indication of recurrent glomerulonephritis is the development of asymptomatic haematuria or proteinuria which may be followed by nephrotic syndrome. Nephrotic range proteinuria occurs in up to 30% of patients after transplantation with the most common cause being transplant glomerulonephropathy.

Focal segmental glomerular sclerosis is the most troublesome recurrent disease, occurring early and leading to premature graft failure. The remaining forms of glomerulonephritis are associated with either minor urine abnormalities or a slow course of transplant dysfunction. Systemic activity of primary diseases such as antiglomerular basement membrane

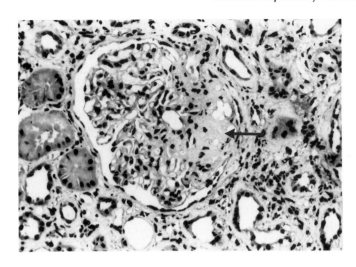

Figure 11.13 Recurrent focal sclerosing glomerulonephritis four months after renal transplantation. The recipient, a 19 year old male, had heavy proteinuria which continued with deteriorating renal function until time of transplant nephrectomy 12 months after transplantation. An area of focal sclerosis is arrowed (H&E,x80).

Table 11.3   Recurrent primary disease in renal transplants

Type	Approximate frequency	Clinical	Significance
Focal segmenting glomerulosclerosis	30–70%	Early proteinuria and graft loss	Poor outcome
Membranous nephropathy	10%	Initial heavy proteinuria	Gradual graft dysfunction
Mesangiocapillary type I	20%	Microsopic haematuria and proteinuria	Shortened graft survival
Mesangiocapillary type II	95%	Minor proteinuria	Almost unimportant
IgA nephropathy	50%	Minor proteinuria	Almost unimportant
Schölein-Henoch purpura	80%	Minor proteinuria	Possible graft failure
Anti-GBM nephritis	5% (if recipients have low anti-AGM antibody	Minor proteinuria or haematuria levels)	Rare graft failure
Diabetes	Probably 100% without pancreas transplant	Proteinuria Hypertension	Good prognosis to at least 10 years
SLE	1%	Minor	
Oxalosis	>90%	Early graft failure	Consider combined hepatic transplantation
Amyloidosis	30%	Proteinuria	Shortened graft survival

nephritis, Schönlein-Henoch purpura and systemic lupus erythematosus should be allowed to subside before transplantation is considered. Diabetes is the commonest systemic condition causing endstage renal failure but recurrent diabetic nephropathy is an uncommon cause of graft failure in the first ten years after transplantation. No primary renal disease provides an absolute contraindication to transplantation on at least one occasion.

The use of cyclosporin for a range of primary renal diseases led to speculation that the incidence of recurrent glomerulonephritis may be less in cyclosporin treated renal transplants. This optimism is probably misplaced.

Immunosuppression withdrawal and non-compliance

The influence of non-compliance with immunosuppression on long term graft survival is difficult to quantitate, but may contribute up to 25% of graft losses in the long term. Factors commonly associated with poor compliance are listed in Table 11.4. The most widely discussed group of patients are adolescents, who for reasons of peer group pressure, perceived social inadequacy and perhaps lack of insight may stop immunosuppression abruptly. Although they often do so without prior warning, when patients in this age group complain about their outward appearance, in particular acne, altered facial features and hirsutism, serious thought must be given to the possibility. Frequent and frank discussion on ways and means of alleviating the symptoms and signs is warranted.

After abrupt cessation of immunosuppression, renal transplant recipients usually present one to four weeks later with an uncomfortable swollen renal transplant, oliguria, weight gain and serum creatinine between 300 and 500 μmol/l. Needle core biopsy of the transplant will demonstrate acute rejection with a predominant cellular infiltrate that may respond to intravenous bolus injections of methylprednisolone. Even with effective reversal of rejection there is, however, inevitable loss of nephrons and the GFR will not return to previous levels. If the renal transplant is lost as a result of acute rejection, the transplant clinician is left with a difficult decision of whether or not to place the patient back on the waiting list for retransplantation. Our experience is that a period of dialysis serves as a salutory reminder of the benefits of renal transplantation and changes the

Table 11.4   Factors associated with poor compliance with immunosuppression

1.	Dialysis non-compliance
2.	Non-attendance at transplant follow-up clinics
3.	Long distances between transplant clinic and home
4.	Paediatric and adolescent recipients
5.	Racial minority recipients
6.	Language difficulties

attitude of previously non-compliant patients.

The more difficult group of non-compliant patients are those that take their imnmunosuppressive drugs on an irregular basis, often loading themselves with cyclosporin prior to a transplant clinic appointment, and giving the appearance of compliance when cyclosporin levels are performed. These patients are often also non-compliant with their other medication and have poorly controlled blood pressure. Their presentation is usually with gradually deteriorating renal function, and their transplant, when biopsied, demonstrates a mixture of changes attributable to acute and chronic rejection. They will often deny non-compliance, but a history from family members, details of supplies of medications at home or infrequent requests for medication prescriptions may point to non-compliance.

Development of immune tolerance to a renal transplant is uncommon, but nevertheless, anecdotal and isolated reports exist of prolonged renal function despite cessation of immunosuppression. Immunosuppression may be withdrawn deliberately in an attempt to control post-transplant lymphoproliferative disorders. These patients are often profoundly immuno-suppressed as a result of prolonged periods of treatment with anti-lymphocyte preparations, and transplant function may continue for two or three months after cessation of oral immunosuppressive therapy. Stopping immunosuppression for other patients with malignant tumours, in the hope that their tumour may regress, is usually unrewarding. It is often better to maintain renal function by continuing immunosuppression, and to provide symptomatic relief for tumour associated problems, than hope to control the tumour better in a dialysis dependent patient.

Integrated management of renal transplant dysfunction

Management of late graft dysfunction implies that the patient has been monitored sufficiently well to detect changes in graft function at a stage when the processes are reversible. Patients do not usually complain of any symptoms and unless an accurate measure of GFR is monitored much of the kidney may be destroyed before changes in serum creatinine are noted. Hypertension, urinary tract infections, local symptoms or changes in urine analysis such as the onset of proteinuria may, in some patients, provide a stimulus to closer investigation.

The key steps in investigation are transplant ultrasound examination or transplant biopsy or both. Unfortunately, in many instances the ultrasound examination will be normal and the biopsy will demonstrate non-specific vascular, glomerular and tubular changes. The transplant clinician will have the options of decreasing the cyclosporin dose to reduce possible related nephrotoxicity, increasing maintenance immunosuppression to slow down the presumed insidious loss of graft function as a result of immunologic processes, or doing nothing. All too frequently, this choice is made without objective evidence to support one option or the other. It is of paramount importance to maintain the function of remaining nephrons for as long as possible, by providing optimum blood pressure control and restricting dietary protein. The time course to eventual graft failure is variable and is best monitored by measurement of GFR rather than serum creatinine. A suggested management pathway for the investigation and treatment of late renal transplant dysfunction is illustrated in Figure 11.14.

Management of the failed graft

The point at which a renal transplant can be said to have failed is based on the patient's symptoms of uraemia, serum biochemistry and perhaps a transplant needle core biopsy. In this respect, it is no different to the decision making process when the patient first presented for renal replacement therapy. The options for the patient with a failing graft are early retransplantation, dialysis, or in some circumstances withdrawal of treatment. In most instances, the gradual decline in transplant renal function will have been followed in a longitudinal fashion, allowing adequate and informed choice between these three treatment options.

Withdrawal of immunosuppression

If early retransplantation is possible, either because of the availability of a living related donor or the possibility of an offer of a cadaver kidney at short notice, it is advisable to continue with maintenance immunosuppression. Otherwise, an attempt can be made to withdraw immunosuppression by stopping cyclosporin and azathioprine, combined with gradual reduction of prednisolone by 1 mg per month. An acute rejection episode may follow within days to weeks in about half the patients and is an indication for graft nephrectomy. In others, however, especially those with small fibrotic kidneys that have been in place for many years, no further problems will be encountered, and the kidney will continue to involute. The limiting factor for retransplantation is usually the degree of sensitisation to histocompatability locus antigens. It is still unclear if removal of the chronically rejected renal transplant will be associated with a decrease in the degree of sensitisation, but there are advocates for routine nephrectomy.

Late transplant nephrectomy

The decision to abandon a renal transplant is a finely balanced one, but should be undertaken if the advantages outweigh the risk associated with an additional difficult surgical procedure. The possible indications for late graft nephrectomy are listed in Table 11.5, the most common of which is an acute exacerbation of rejection during or after withdrawal of maintenance immunosuppression. The surgical procedure is made difficult by the extensive fibrous reaction to the presence of the renal transplant, a problem which can be overcome by intracapsular dissection of the renal transplant. The procedure nevertheless is one for an experienced surgeon.

Prior knowledge of the position of the hilum of the transplant and the renal vessels is essential before embarking upon a late transplant nephrectomy.

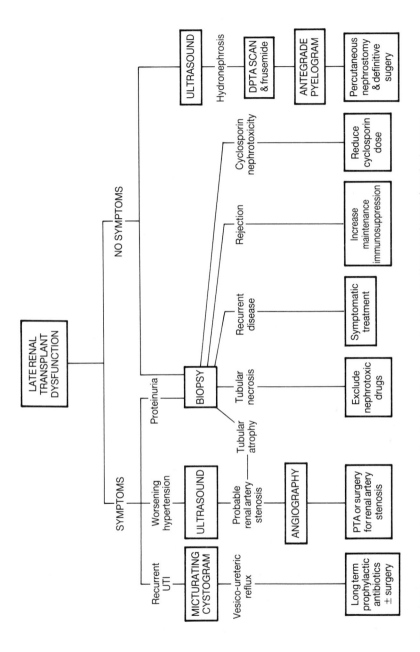

Figure 11.14   Integrated management plan for investigation and treatment of late renal transplant dysfunction.

Table 11.5   Indications for late transplant nephrectomy

1. Chronic rejection and failed withdrawal of immunosuppression
2. To provide for retransplantation
3. To possibly reduce histocompatibility locus antigen sensitisation
4. Chronic transplant infection (for example, tuberculosis)
5. Recurrent UTI and transplant vesico-ureteric reflux in the failed graft
6. Recurrent primary renal disease and uncontrolled heavy proteinuria
8. De novo renal transplant malignancy

This information can be obtained from the description of the original surgical procedure or from an ultrasound examination. The old transplant wound is re-opened and the capsule of the renal transplant is opened. Intracapsular blunt dissection is then undertaken down to the hilum of the kidney, at which point the capsule is crossed again to allow dissection of the renal artery branches and veins in the region of the hilum of the kidney. On some occasions, it may be possible to identify the anastomosis to the recipient vessels. However, this is usually difficult and unnecessary. The ureter is usually more easily identified and can be ligated at point of entry to the bladder. Careful haemostasis is obtained by diathermy to the multiple capsular bleeding vessels. A large suction drain is placed into the intracapsular cavity created by the removal of the kidney and the wound closed. An inpatient stay of four to seven days can be expected.

Psychological support

The decision to return to dialysis is taken surprisingly well by most renal transplant recipients, perhaps because of the transplant clinic environment in which they have had the opportunity to observe and discuss similar problems encountered by other transplant recipients. Nevertheless, the loss of renal transplant function and the need to return to dialysis is inevitably associated with loss of self esteem and reduced employability. For this reason, any decision to return a transplant patient to dialysis should be taken in full consultation with the patient and his or her family, and appropriate social support provided as required. The possibility of retransplantation and return to more normal health should be seen as a realistic goal in the majority of patients. This may provide a degree of psychological support to the patient as he or she returns to dialysis.

Further reading

1. Matas AJ, Brayman KL, Gillingham KJ. Presentation of renal transplant graft survival data – where should 'death with a functioning graft' be included? *Clin Transplant* 1990; 4, 142–144.
2. Tilney NL, Whitley WD, Diamond JR *et al.* Chronic rejection – an undefined conundrum. *Transplantation* 1991; **52**, 389–398.
3. Perico N, Remuzzi G. Cyclosporin induced renal dysfunction in experimental animals and humans. *Transplant Rev* 1991; **5**, 63–80.

4.   Almond PS, Gillingham KJ, Sibley R *et al*. Renal transplant function after 10 years of cyclosporin. *Transplantation* 1992; **53**, 316–323.
5.   DeMeyer M, Pirson Y, Dautreband J *et al*. Treatment of renal graft artery stenosis. *Transplantation* 1989; **47**, 784–788.
6.   Kinnaert P, Hall M, Janssen F *et al*. Ureteral stenosis after kidney transplantation; true incidence and long term follow up after surgical correction. *J Urol* 1985; **133**, 17–20.
7.   Mathew TH. Recurrent disease after renal transplantation. *Transplant Rev* 1991; **5**, 31–45.
8.   Chiverton SG, Murie JA, Allen RD *et al*. Renal transplant nephrectomy. *SGO* 1987; **164**, 324–328.

12 Imaging the renal transplant

Aims of imaging

The aim of all forms of imaging is to gain an understanding of pathophysiological processes taking place in the graft, without recourse to invasive techniques such as biopsy or operative intervention. It is possible to divide the situations in which imaging may be used into routine investigations designed to monitor aspects of graft function, and diagnostic investigations undertaken to determine the cause of a change, such as decline in renal function. Against the advantages of gaining this information have to be set the disadvantages and risks of each test. Repeated use of ultrasound has no known complication while repeated use of intravenous radiocontrast has a known effect on renal function and the potential to produce dramatic anaphylaxis.

Routine imaging

There are a number of investigations where serial measures of a patient's graft provide greater amounts of information than can be achieved with a single test. The ultrasound measure of graft size and arterial blood flow velocity are examples where changes in the data with time provide more information than a single test. Swelling of the kidney and reduction in diastolic blood flow velocity may provide useful evidence in favour of acute rejection, while single measurements are insensitive. Accurate assessments of the glomerular filtration rate (GFR) using radiolabelled EDTA or DTPA yield clear information on whether or not a graft is deteriorating over a period of months or years, but requires regular measurements to be made. Routine assessment of a graft by ^{99m}TcDTPA scanning may be warranted after transplantation of a graft with multiple arteries, to ensure that adequate perfusion of all parts of the kidney has been achieved.

Diagnostic imaging

The problem that most frequently confronts those caring for transplant recipients is that of declining renal function. As described in Chapters 9 and 10 urgent and accurate investigation is required to direct treatment or determine operative intervention. Imaging techniques should be applied in such a way as to minimise the risk and expedite the diagnosis. Table 12.1 lists the commonly available techniques with some of their particular strengths and weaknesses.

Ultrasound

Advantages

Ultrasonography is entirely non-invasive; the equipment is portable on purpose built trolleys so it may be brought to the patient's bed side or even anaesthetic room in the operating suite; examination is relatively rapid and results are available immediately; images may be recorded and stored in a variety of ways; accurate anatomical location may be made in relation to the skin surface; accurate measurements of size, volume and blood flow velocities can be made and assessments of perirenal structures and bladder are routinely available.

Disadvantages

Operator skills are required to create acceptable and reproducible images; resolution of small structures is poor; obese people are difficult to assess accurately; and bowel gas may prevent examination of important structures.

Table 12.1 Advantages and disadvantages for alternative techniques for imaging renal transplants

Technique	Advantages	Disadvantages
Ultrasound	Non-invasive Repeatable	Cannot assess function
Radionuclide scanning	Low radiation dose Assesses function	Limited repeatability
Plain radiology	Simple	Limited diagnostic information
Intravenous pyelography	Good anatomical definition	Risk of contrast nephrotoxicity Poor in poorly functioning grafts
Angiography	Definitive for vascular pathology	Invasive Risk of contrast nephrotoxicity
Computerised tomography	Good anatomical information	Cannot assess function Expensive

Sector scanning

It is routine practice to assess the size and surface of a renal transplant using a scanning ultrasound probe of 3–5 mHz. This provides a two dimensional picture of the graft and allows measurement of the length and width of the kidney. These measurements are subject to minor variation so that, for reproducible results, careful orientation of the kidney in the field of view is essential, with sagittal views in the long axis plane of the graft and transverse views at 90° to the long axis. Calculation of the volume of the kidney has been used to show swelling, which may be associated with acute rejection, but is difficult to estimate reliably. The surface of the kidney should be examined for scars or areas of damage and the echo pattern of the renal tissue assessed. In rejecting grafts there may be decreased corticomedullary differentiation but this is often a late finding and not sufficiently clearcut to be useful to the clinician.

Imaging of the renal sinus, containing the collecting system and perihilar fat, may provide evidence of pelvicalyceal dilatation and be the first indicator of partial ureteric obstruction (Figure 12.1). It may be possible to trace the ureter towards the bladder when there is obstruction, but normally it is not easily visible.

Perirenal tissues may be examined for the presence of haematoma or other fluid collections such as lymphoceles. Wound collections are most clearly defined by ultrasound, with the advantage of being able to define the tissue planes which restrict the collection. Other pelvic organs such as the uterus and ovaries can be included within the standard examination. The bladder should be visible to ultrasound unless drained by a urinary catheter or recent voiding. It may be possible to see blood clot within the bladder and in male patients to assess the prostate.

Accurate localisation of the kidney may be used to direct a biopsy needle, either by surface marking or by using a needle biopsy guide attached to the ultrasound probe. This is especially useful in obese patients and when bowel overlies the kidney.

Doppler ultrasound

Using the principle of Doppler shift of the wavelength of ultrasound reflected from moving blood, it has proved possible to visualise renal transplant arterial and venous blood flow. The first reports of this technique in transplants were tinged with optimism, while subsequent reports have been mellowed by wider experience.

Arterial blood flow Doppler signals are normally characterised by an abrupt systolic peak velocity followed by a diastolic phase of slower blood flow (Figure 12.2). A number of attempts have been made to provide a single number which describes this waveform including; the diastolic to systolic ratio; pulsatility index; area under the curve for half the cardiac cycle; and the Pourcelot ratio. The most popular of these is the Pourcelot ratio or Resistive Index (RI) which is calculated according to the simple equation detailed in Figure 12.2. All are essentially measures of the resistance to blood flow in the renal vascular bed. A high vascular resistance prevents forward blood flow during diastole, reducing the diastolic component of the

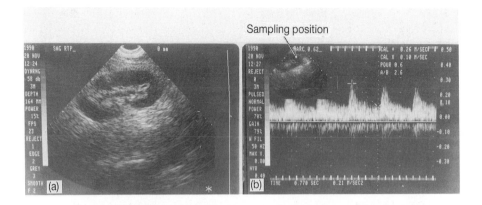

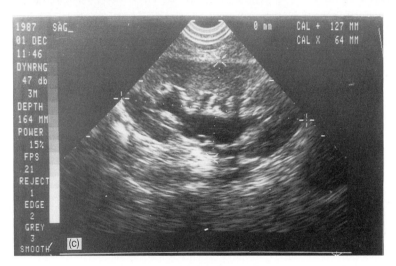

Figure 12.1 (a) Renal transplant ultrasound demonstrating the outline of the kidney and normal renal sinus with no evidence of hydronephrosis. (b) Duplex ultrasound scan of the arcuate arteries. The sampling position is shown from the small white square in the top left hand image of the renal transplant (arrow). The peak systolic and end diastolic flows are marked with the calculation of the Pourcelot ratio (POUR) = 0.6 which is shown in the upper right hand corner of the screen. (c) Renal ultrasound sector scan demonstrating renal size (127mm × 64mm) and both renal pelvis and calyceal dilatation. The axes of the kidney are marked with crosses for distance measures.

waveform (Figure 12.3). In the extreme case, such as occurs with complete renal vein thrombosis, blood that flowed into the arterial tree during systole may reverse during diastole and actually flow back down the renal artery (Figure 12.4).

Changes in the Doppler pattern may be induced by a number of different pathological conditions. The major influences on the RI are detailed in Table 12.2, from which it is clear that the sensitivity and specificity of the RI for diagnosis of acute rejection, especially acute cellular rejection,

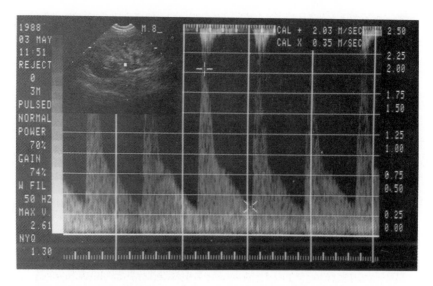

Figure 12.2 Renal Doppler ultrasound. Arterial Doppler frequency trace with cross marks on peak systolic flow and end diastolic flow rate. Pourcelot ratio (resistive index) calculated by the equation

RI = peak systole (2.03) – end diastole 0.35/peak systole (2.03)
RI = 2.03 – 0.35/2.03 = 0.828.

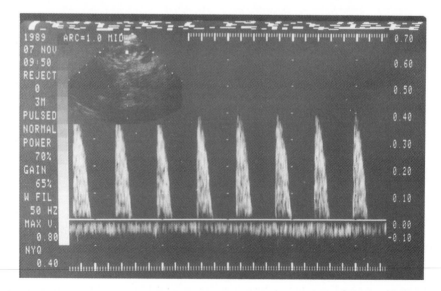

Figure 12.3 Renal transplant Doppler ultrasound. Arcuate vessel Doppler assessment of blood flow RI = 1.0. The Doppler ultrasound demonstrates systolic flow only, with absent diastolic flow. Below the baseline is the continuous signal of nearby venous flow. Simultaneous renal biopsy demonstrated severe acute vascular rejection.

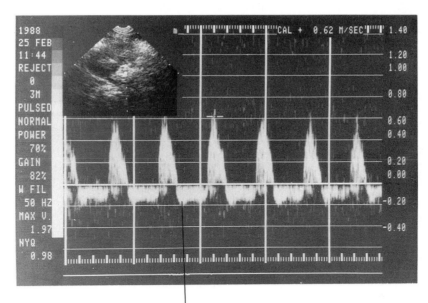

Reversed diastolic blood flow

Figure 12.4 Renal transplant Doppler ultrasound demonstrating reversed diastolic blood flow. In this ultrasound the peak systolic flow is followed by reversed diastolic blood flow during systole. The reversed flow is in the opposite direction and thus appears below the baseline (arrow). Simultaneous biopsy demonstrated severe cellular and vascular rejection.

is low. Like many of the tests that are used in clinical medicine, the results of the Doppler ultrasound are more useful if they are measured serially and considered in the context of both the clinical picture and the complete ultrasound examination.

Absence of detectable renal blood flow by Doppler ultrasound may either be due to difficulty with the examination, for technical reasons, or to loss of renal blood flow. Colour Doppler demonstration of the kidney may be helpful in distinguishing these situations, since arterial and venous flow in the renal parenchyma can be hard to detect using conventional Doppler images. If blood flow to the graft has been lost it is either a surgical emergency in which minutes matter or indication for planned graft nephrectomy. Warm ischaemia of the kidney, induced by arterial kinking or thrombosis of the artery or vein, can only be tolerated for very short periods. If there is reason to believe the graft has very recently ceased to function, confirmation of the Doppler ultrasound finding should be at open exploration within minutes. If, on the other hand, graft function has been absent for some time it would be reasonable to confirm the finding of loss of blood flow by DTPA scanning before planned nephrectomy.

Doppler ultrasound can be useful to the clinician in a number of ways to investigate specific situations:

Initial non-function of the graft   Doppler ultrasound confirms that blood

Table 12.2   Pathophysiological correlation with Doppler ultrasound

Doppler ultrasound RI	Likely pathology*	
0.65–0.80	1	Normal
	2	Acute tubular necrosis
	3	Acute rejection
	4	Ureteric obstruction
0.80–0.95	1	Acute tubular necrosis
	2	Acute rejection
	3	Ureteric obstruction
	4	Normal
	5	Venous obstruction
0.95–1.0	1	Acute rejection
	2	Acute tubular necrosis
	3	Ureteric obstruction
	4	Vascular obstruction
Sequential rise in RI	1	Acute rejection
	2	Ureteric obstruction
	3	ATN if the first result was day 1 or 2
Reversed diastolic flow	1	Severe acute rejection
	2	Renal vein obstruction
	3	Ureteric obstruction
Absent signal	1	Vascular thrombosis
	2	Ultrasound machinery malfunction
	3	Operator malfunction
	4	Patient difficult to examine
Miscellaneous findings		
Low RI < 0.6	Dopamine infusion	
Turbulent venous flow	Arteriovenous fistula	
Rapid peak systolic flow in main artery	Renal artery stenosis	

* Likely renal pathology is in approximate order of frequency

is flowing in the kidney. The differential diagnosis is between hyperacute or accelerated rejection and acute tubular necrosis. The RI is not helpful and may mislead. In the immediate postoperative period, especially when dopamine is used, the RI may be in the range of 0.65–0.80 with ATN, and then rise to 1.0 over the next 24–48 hours. On the other hand the RI may be 1.0 because of severe rejection. It is not safe to rely upon the RI to help in the differential diagnosis, however, if it is in the range 0.65–0.80, it is unlikely that the initial non-function is due to rejection.

Monitoring acute tubular necrosis   Each graft has an individual blood flow and microvascular resistance. Therefore, when the RI has been established for a given kidney by serial measurements, it is possible to watch the evolution of tubular necrosis. The normal pattern is for the RI to drop slowly as ATN recovers, predating both the flow of urine and falling serum creatinine. If acute rejection intervenes in a graft suffering ATN, the first indicator may be deterioration in the RI. Regular measures of RI in this situation may thus help to decide the timing of renal biopsy.

Acute decline in graft function   A rise in RI in a graft which has acutely deteriorating function suggests acute rejection if the ultrasound examination excludes ureteric obstruction. A rise in RI to a value of 1.0, implying loss of end-diastolic blood flow, has a poor prognosis unless due to ureteric obstruction or acute tubular necrosis, because it implies severe vascular or cellular rejection.

Defining the cause of a renal bruit   While soft renal bruits may often be heard in the first hours after transplantation, a bruit that appears later may be due to renal artery stenosis or an arteriovenous fistula. Colour Doppler ultrasound is extremely accurate for detecting and localising an arteriovenous fistula because of disturbance in the normal venous flow patterns. Renal artery stenosis may be detected by Doppler ultrasound from the high velocity blood flow through the stenosed segment. Whether Doppler ultrasound can exclude significant renal artery stenosis or not depends upon being able to examine the full length of the main renal artery. Since this is often not feasible, the angiogram remains infinitely superior for this purpose.

Radionuclide studies

Renal imaging using radionuclide scanning provides a low risk assessment of blood flow and function together with limited anatomical information. A variety of agents can be used depending upon the focus of interest (Table 12.3). Both EDTA and DTPA are freely filtered by the glomerulus and are neither absorbed nor metabolised in the tubule. These agents are thus useful for measurement of glomerular filtration rate. DMSA, on the other hand, is best utilised to discern areas of poor function within a kidney, such as that occurring with pyelonephritic scarring, since it is excreted almost exclusively by tubular function. ^{99m}Tc DTPA and ^{99m}Tc DMSA provide a dose of between 28 and 46 mrad to the gonads, which compares favourably with

Table 12.3   Radionuclides most frequently used in renal transplantation

Agent	Main uses
^{51}Cr EDTA	Measurement of glomerular filtration rate
^{99m}Tc DTPA	Measurement of glomerular filtration rate
	Renal perfusion (index)
	Renal function assessment
	Identification of urinary leakages
^{131}I-hippuran	Measurement of effective renal plasma flow
	Renal function assessment
^{99m}Tc DMSA	Assessment of functioning renal parenchyma
^{99m}Tc Gluconate	Assessment of functioning renal parenchyma
^{99m}Tc MDP	Bone scan
^{67}Gallium Citrate	Investigation of occult infection
EDTA	Ethylenediamine tetra-acetic acid
DMSA	2,3–dimercaptosuccinic acid
DTPA	Diethylene triamine penta-acetic acid
MDP	Methylene diphosphonate

the 640–800 mrads from the standard intravenous pyelogram series using X-rays.

DTPA Scanning

⁹⁹ᵐTc DTPA is used frequently to provide images of renal transplant blood flow, function and urine flow. Serial scans of a poorly functioning graft can help in diagnosing changes such as acute rejection occurring in a transplant suffering from acute tubular necrosis. Using a gamma camera, images taken at one second intervals document the flow of blood to the kidney and its perfusion. Subsequent images taken over the next 30 minutes, or in some circumstances longer, demonstrate renal function and flow of urine to the bladder (Figure 12.5).

Acute tubular necrosis   Renal perfusion may be mildly to moderately reduced although there is tracer uptake, there is no concentration of tracer in the kidney and little or none in the bladder (Figure 12.6a and b). The DTPA that appears in the kidney washes out progressively back into the intra- and extravascular space. If repeated scans are performed

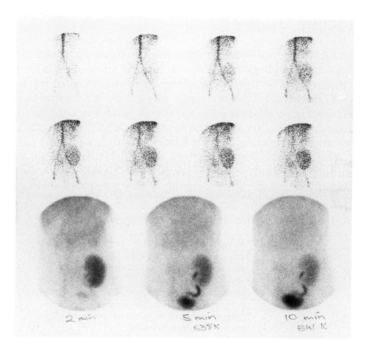

Figure 12.5   ⁹⁹ᵐTc-DTPA scan. Normal scan. In the top left hand picture radionuclide is seen to be passing down the aorta to the level of the bifurcation. In the second figure radionuclide is seen to flow to the femorals and into the patient's left sided renal allograft. In the subsequent pictures, flow to the renal transplant is seen to increase progressively with the one second images. The lower series of pictures taken at 2 mins, 5 mins, and 10 mins demonstrate concentration of ⁹⁹ᵐTc DTPA with excretion through the ureter to the bladder.

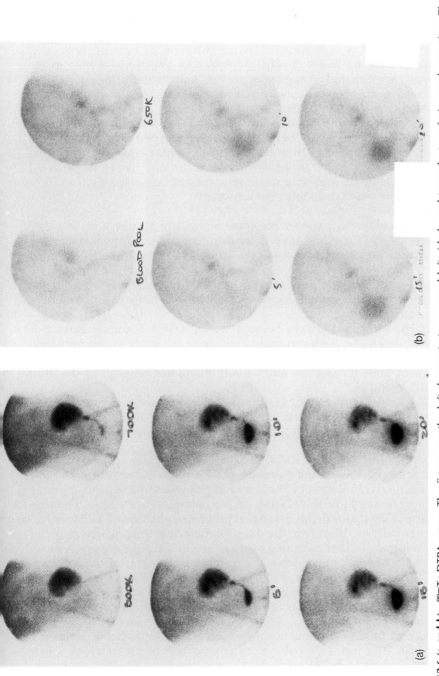

Figure 12.6 (a and b) ⁹⁹ᵐTc-DTPA scans. The figures on the left demonstrate a normal left sided renal transplant perfusion and excretion. The figure on the right is a case of acute tubular necrosis with poor uptake into the right sided renal transplant and slow excretion of a small quantity of radionuclide into the bladder.

during the evolution of ATN, perfusion improves and the kidney starts to concentrate DTPA.

Acute rejection   Renal function declines with acute rejection and with it the DTPA perfusion is reduced and concentration of radionuclide poor. Some centres use an index of renal perfusion in which blood flow to the kidney can be estimated. A region of interest may be defined over the iliac artery distal to the transplant, with a second region defined over the graft itself, excluding the iliac artery. A comparison can then be made of the relative blood flow to the kidney and to the leg. The ratio of the areas under the curve of these two regions of interest provides the perfusion index. A rise in this index has been shown to be associated more frequently with acute rejection than with ATN. To be useful this technique has to be repeated serially and routinely and may result in a patient being scanned multiple times in the first few weeks after the transplant.

Vascular anatomy   In patients transplanted with a kidney with multiple renal arteries, DTPA scans may demonstrate regional differences in the kidney. At the extreme, when one of the arteries has thrombosed, a segment of kidney may be shown to be without a blood supply. In some instances, when cold perfusion and preservation has been worse through one of the arteries, one may see a normal segment of kidney side by side with a segment demonstrating ATN.

Anuric grafts should always be assessed by DTPA scanning since it is possible to determine with certainty whether or not there is still perfusion of a potentially viable graft. A 'white hole' where the kidney should be is indicative of graft thrombosis and usually mandates removal of the graft (Figure 12.7a and b).

Renal artery stenosis and cyclosporin nephrotoxicity cannot be adequately diagnosed by radionuclide imaging alone. Both cause a decrease in renal function and the scan appearance may be indistinguishable from that of rejection.

Ureteric anatomy   Following clearance of DTPA into the urine it is possible to see steady accumulation of the tracer in the collecting system, passage down the ureter and into the bladder or ileal conduit (Figure 12.8). The scan is particularly useful for diagnosis of two problems: obstruction and leakage. Ureteric obstruction, usually in the distal portion or at the junction of ureter and bladder, may be demonstrated by dilatation of the proximal ureter, slow passage to the bladder and delayed emptying of the pelvicalyceal system (Figure 12.9). When there is doubt about the possibility of obstruction, frusemide may be used to establish a flow of urine of more than 3 or 4 ml/min. This should lead to rapid washout of tracer, so delay may be interpreted as obstruction.

The second urinary abnormality for which DTPA scans prove useful is a ureteric leak. The leak may come from any portion of the pelvis or ureter but usually occurs in the distal portion five to ten days after the operation. The precarious blood supply to the ureter ensures that the distal ureter is most vulnerable to necrosis and thus to leakage of urine. It is not usually possible to arrive at an unequivocal anatomical localisation of a leak, but

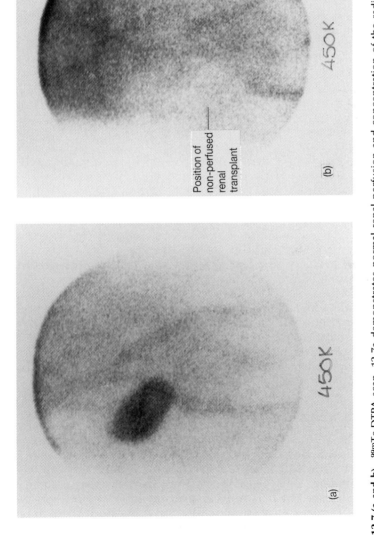

Figure 12.7 (a and b) ⁹⁹ᵐTc-DTPA scan. 12.7a demonstrates normal renal perfusion and concentration of the radionuclide. 12.7b demonstrates a 'white hole' (arrow) with no renal perfusion in a case of renal artery thrombosis. These two pictures come from the same patient and were taken two days apart. At exploration the artery was found to be obstructed by embolus, which was presumed to have come from either the left atrium or left ventricle.

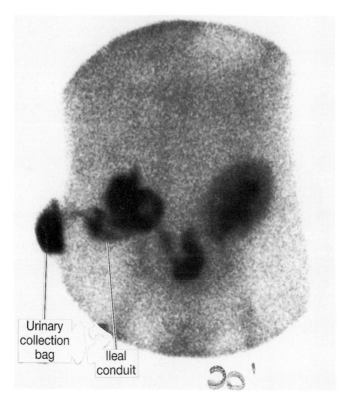

Figure 12.8 99mTc-DTPA scan of a renal transplant drained through an ileal conduit to a right sided urinary collection bag (film taken at 20 mins).

unequivocal evidence of a leak is useful as a precursor of a return to the operating room (Figure 12.10a and b).

Lymphocoele   Any sizable fluid collecting around the kidney such as a lymphocele or haematoma may be detected using DTPA. Fluid collections, which are not of course perfused with blood, appear as 'holes' in the later images (Figure 12.11). Ultrasound and computerised tomography however provide more reliable anatomical localisation, with the former being the investigation of choice.

DMSA scanning

99mTc DMSA is localised in the proximal tubular cells and thus provides images of functioning cortical tissue (Figure 12.12). An alternative preparation is 99mTc Gluconate since it is handled in a similar way but has greater urinary excretion. It is possible to detect cortical changes in many well functioning grafts but it is not yet clear whether or not these represent sequels to vascular injury sustained by the graft and if these appearances are predictive of graft outcome. 99mTc DMSA is useful, on occasions, to confirm graft pyelonephritis.

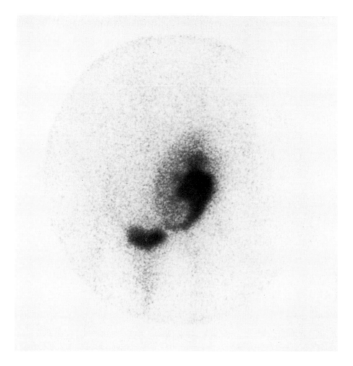

Figure 12.9 99mTc-DTPA scan. 20 min film. In this film taken 20 mins after injection, radionuclide has concentrated and excreted into the dilated pelvicalyceal system, relatively small quantities having passed through the lower ureteric obstruction to the bladder. Ureteric obstruction was diagnosed and surgically corrected with anastomosis of the left native ureter to the dilated pelvis of the renal transplant.

Glomerular filtration rate

Serum creatinine is used as a reliable indicator of renal transplant function and has proved to be sensitive in the short term when measured serially using a single well calibrated analyser. In the long term it is, however, not a sensitive indicator of chronic damage to a relatively well functioning graft, especially when the GFR is greater than 35 ml/min/1.73m². Glomerular filtration rate may be estimated from creatinine clearance but the inconsistencies, particularly in collection of a full and accurately timed 24 hour urine, make it unreliable in the routine clinic.

The alternative used by transplant units is routine measurement of GFR using elimination of a radionuclide. Either 51Cr EDTA or 99mTc DTPA may be injected intravenously and blood samples taken at least two or three times over the subsequent four hours. When renal function is good, a reasonable estimate can be made from a single timed sample, but this is not sufficiently reliable for transplant GFR. Annual measurement of GFR provides the best indication of chronic change in graft function. Correction for surface area is especially relevant in transplant patients whose surface area may fluctuate due to weight changes induced by corticosteroids and overeating.

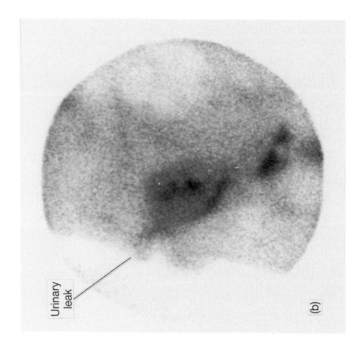

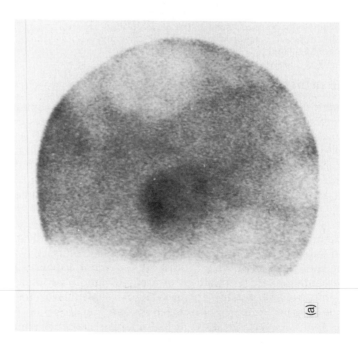

Figure 12.10 (a and b) ^{99m}Tc-DTPA scan of renal transplant. In 12.10a, contrast is seen to collect in a poorly functioning renal transplant. In the 20 minute scan in 12.10b, contrast is seen to be concentrated in the pelvicalyceal system passing to the centrally placed bladder but also onto the surface of the skin (arrow) demonstrating a urinary leak from a necrosed mid ureter.

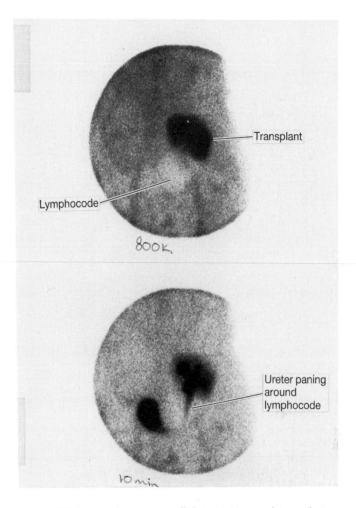

Figure 12.11 ⁹⁹ᵐTc-DTPA scan showing a well functioning renal transplant concentrating the DTPA, beneath which is a white area of absent signal associated with a lymphocele. In the lower figure the ureter can be seen to pass around the lateral and lower portion of the lymphocele (arrow).

Other imaging techniques in renal transplant patients

Radionuclide scanning has two other major applications in renal transplantation. Investigation of renal bone disease, avascular necrosis and osteomyelitis all use ⁹⁹ᵐTc MDP to provide an image of bone perfusion and uptake of tracer into bone. Renal osteodystrophy may be seen as a diffuse increase in uptake, while in the early phases of avascular necrosis, focal areas of decreased uptake occur in relation usually to the femoral heads. As avascular necrosis of bone develops a hot spot may appear with progression to healing. Hot spots on the bone scan may yield evidence for

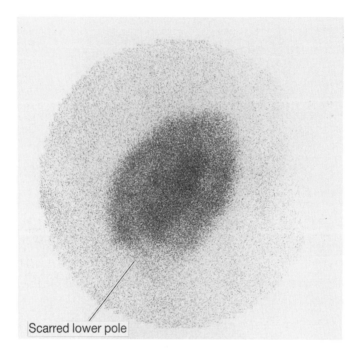

Scarred lower pole

Figure 12.12 99mTc-DMSA scan. The renal outline is clearly seen with a segment of the lower pole scarred due to pyelonephritis (arrow).

osteomyelitis or fracture without distinguishing between these alternative diagnoses (Figure 12.13).

Radionuclide cystograms using 99mTc sulphur colloid may substitute for radiocontrast studies. Detection of ureteric reflux in both native and transplant ureters is possible using either technique. Conventional X-rays have the disadvantage of providing a larger radiation dose but the advantage of accurate anatomical pictures.

Gallium and radiolabelled leucocytes have both been used to localise areas of infection in patients suspected of having abscesses, but without features that allow anatomical localisation. While these investigations may prove useful in occasional patients they are usually least helpful when there is most doubt and confirm findings in situations that are clear.

Nuclear medical techniques, of course, have widespread use in normal investigation and management of co-existing disease in renal transplant recipients. Cardiac assessment using both the gated heart pool scan and stress thallium perfusion scan, together with ventilation and perfusion scans of the lungs, comprise the majority of these other extrarenal tests.

Radiology

Conventional X-rays are becoming steadily less important in evaluation

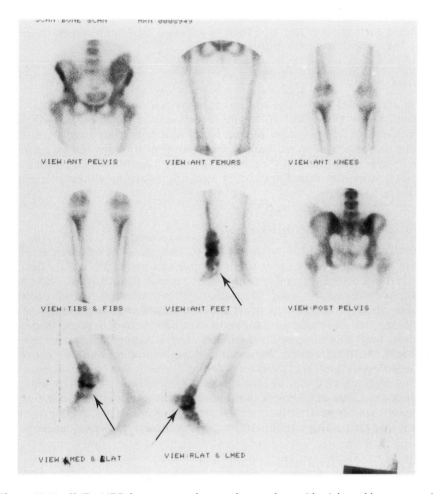

VIEW:ANT PELVIS   VIEW:ANT FEMURS   VIEW:ANT KNEES

VIEW:TIBS & FIBS   VIEW:ANT FEET   VIEW:POST PELVIS

VIEW:MED & LLAT   VIEW:RLAT & LMED

Figure 12.13 99mTc MDP bone scan of a renal transplant with right ankle osteomyelitis. The top left film demonstrates the right sided renal transplant superimposed on the pelvis. Bone uptake reveals concentration in the right ankle demonstrated in the anterior, medial and lateral views (arrows).

of the renal transplant with development of ultrasound and radionuclide scanning techniques. Intravenous pyelography has, for example, been almost entirely superceded and would now provide less than 1% of images of transplants in most programmes. There are, however, instances where there is no replacement for conventional radiological techniques such as arteriography and antegrade pyelography. Contrast media for radiographic procedures contain iodine and are both highly osmolar and charged. Newer agents are the so called 'non-ionic' contrast agents. The incidence of acute nephrotoxicity from these agents is, in well hydrated patients, in the region of 8 to 10%. Experimental evidence that the non-ionic media are less nephrotoxic has not yet been borne out in clinical practice, though they may be less prone to inducing serious allergic effects. The incidence of nephrotoxicity rises in patients with poor renal function, diabetes and myeloma. It is particularly important in these high risk groups to reconsider the need for radiocontrast and then to proceed only with careful attention to hydration and induced diuresis.

Plain X-ray

While abdominal X-rays are useful for many reasons in patients who have received a renal transplant, for example to assess gastro-intestinal problems, there is almost no reason for a plain X-ray of a renal transplant except as a control film when contrast is used, or to confirm the position of a ureteric stent. Renal calculi in the transplant are very uncommon and ultrasound provides more accurate definition of size, shape and position than a plain X-ray.

Intravenous pyelography

Successful pictures of the renal transplant as a result of intravenous injection of contrast medium rely upon several factors. Firstly, graft function must be sufficient to concentrate and excrete the contrast. Secondly, tomography must be used to evaluate the nephrogram and excretory phases. Thirdly, the patient must be cooperative and not grossly obese. There are, in addition, a number of conflicts between protecting renal function and obtaining optimal images. Dehydration and large doses of iodinated contrast improve the result of the examination but are toxic to the transplanted kidney. As mentioned above, for almost all indications, intravenous pyelography can be replaced by ultrasound or nuclear medicine techniques. Clear anatomical definition of the site of obstruction of the transplant ureter in a partially obstructed but reasonably well functioning graft is one remaining absolute indication for intravenous pyelography.

Antegrade pyelography

Surgical intervention to treat ureteric obstruction is best preceded by accurate definition of the anatomy of the ureter and the site of obstruction. Renal function is often too poor to rely upon contrast excreted by the kidney to outline the course of the ureter. Ultrasound will detect hydronephrosis and may be able to identify a dilated ureter, but seldom defines a stricture in sufficient detail to plan surgical intervention. For these reasons direct injection of contrast into the renal pelvis is the preferred technique. Ultrasound or computerised tomography may be used to guide the needle through renal cortex to the pelvis. Measurement of the hydrostatic pressure may give some indication of the degree of obstruction but contrast injection with or without tomography yields the anatomical detail needed to plan intervention.

Antegrade pyelography may be followed by balloon dilatation of a ureteric stricture in the hands of a skilled operator, and may treat obstruction successfully. Two temporary measures may be taken at the time of antegrade pyelography. A percutaneous catheter terminating in a pigtail in the renal pelvis may be used to drain urine temporarily. Alternatively, if a guidewire can be passed through the obstructed region and into the bladder, a double ended stent may be left across the stenosis. If balloon dilatation has been achieved it is particularly useful to leave the double J ureteric stent for a period of weeks before removal by cystoscope.

Retrograde pyelography

As a rule of thumb it is technically impossible to use retrograde pyelography with a transplant ureter. When an antirefluxing procedure has been used to anastomose the transplant ureter to the bladder, it successfully frustrates the cystoscopist's attempts to pass a catheter up the ureter. When there is free reflux of urine, the transplant pelvis and calyceal system can be visualised using a simple cystogram (Figure 12.14).

Cystography

The bladder and urethral outlines may be seen by instillation of radiocontrast

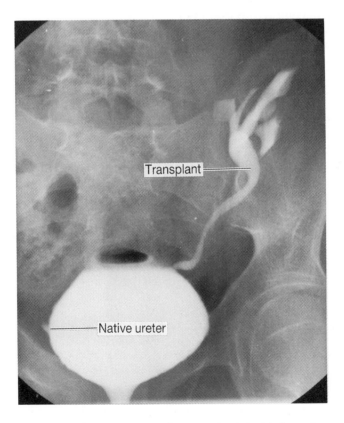

Figure 12.14 Cystogram demonstrating reflux into the left sided renal transplant and minimal reflux into the right native ureter (arrow).

into the bladder through an indwelling catheter. The volume of contrast instilled may be regulated by the surgeon reluctant to test the integrity of a recent ureteric anastomosis; the patient by symptoms of bladder fullness; or the radiologist when an examination has been completed. The structures that can be seen include refluxing native and transplant ureters; cystotomy or anastomotic urinary leakage; and the urethra on micturition. The instilled volume at which bladder discomfort becomes evident should always be noted and may, in long term dialysis dependent patients, be as low as 50 ml. Prophylactic antibiotics should be given before cystography and based upon known bacterial pathogens if present in a recent urine sample. Alternatively a single intravenous or intramuscular dose of a broad spectrum antibiotic such as an aminoglycoside or second generation cephalosporin may be used.

Angiography

Renal transplant angiography is an invasive procedure in which a noxious intravenous contrast agent is infused directly into the kidney. It is thus now an investigation of last resort but it is irreplaceable when the renal artery anatomy is under question, particularly when renal artery stenosis is likely on the basis of clinical, ultrasound or radionuclide criteria (Figure 12.15).

Before examination of a functioning graft the patient should be well hydrated and actively diuresing. Introduction of the angiogram catheter can usually be made via the contralateral femoral artery, though the

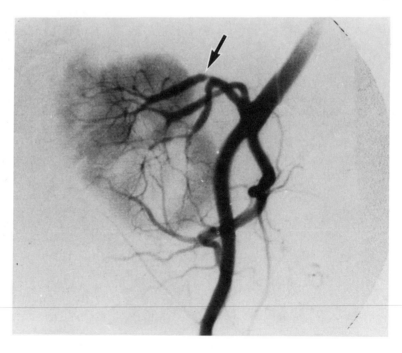

Figure 12.15   Renal transplant angiogram demonstrating early branching of the renal artery with multiple arterial stenoses of the lower branch (arrow).

brachial artery approach may sometimes be needed. The commonest sites for stenosis are the orifice of the renal artery or at 10 to 20 mm from the anastomosis. Arteries transplanted on a patch of aorta do not usually have ostial narrowing. Peripheral pruning of the renal artery tree occurs with acute and chronic graft rejection but neither need now be diagnosed in this way. The major veins can be seen from following the arterial flush through, which may help in diagnosis of renal vein thrombosis. An arteriovenous fistula and occasionally arterio-ureteric fistula can be diagnosed by angiography.

The range of X-ray guided intervention has expanded to include balloon dilatation of transplant renal artery stenosis, though the small but measurable incidence of subsequent thrombosis must be included in the decision to proceed with this therapy. Embolisation of an arteriovenous fistula is seldom justified. Any measures, on the other hand, that stem the flow of blood from vascular to urinary systems will avert nephrectomy as the definitive treatment.

Computerised tomography

There are relatively few indications for computerised tomography of a renal transplant itself. The major advantages are instead in investigation of perirenal structures such as collections of blood clot, pus, urine or lymph. Accurate localisation by ultrasound may not be possible in some situations, e.g. when there is overlying bowel gas. Computerised

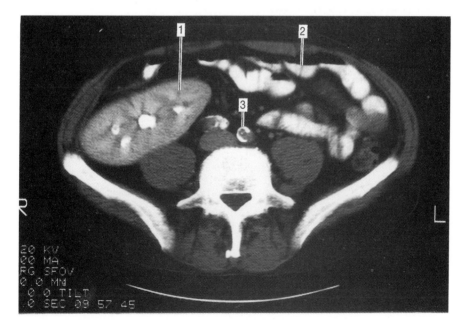

Figure 12.16 CT scan demonstrating a right sided renal transplant excreting intravenous contrast (1). Note also the oral contrast in the bowel with gas contrast fluid levels (2) and the partially calcified wall of the aorta (3).

tomography is most useful when combined with both oral and intravenous contrast. The decision to use the latter must therefore be taken explicitly by the transplant team and be prepared for by hydration and diuresis (Figure 12.16).

Magnetic resonance

Magnetic resonance imaging has been applied to diagnosis of renal transplant dysfunction. The major change visible in the transplant is loss of corticomedullary differentiation. This finding is, however, not specific and may occur with acute tubular necrosis, acute rejection, chronic rejection and cyclosporin toxicity. Therefore, while magnetic resonance is quite sensitive for detecting renal transplant dysfunction it is not helpful in differentiating the cause.

Further reading

1. Maisey ME, Britton KE, Gilday GL (eds). *Clinical Nuclear Medium*. London: Chapman and Hall, 1983.
2. Tauxe WM, Dubovsky EV (eds). *Nuclear Medicine in Clinical Urology and Nephrology*. Norwalk, Connecticut: Appleton-Century-Crofts, 1985.
3. Schwab SJ, Hlatky MA, Pieper KS *et al*. Contrast nephrotoxicity: a randomised controlled trial of a non ionic and an ionic radiographic contrast agent. *New Engl J Med* 1989; **320**, 149–153.
4. Pozniak MA, Kelez F, Dodd GD. Renal transplant ultrasound: imaging and Doppler. *Semin Ultrasound CT MR* 1991; **12**, 319–334.
5. Liou JTS, Lee JKT, Heikan JP, Tolty WG, Molina PL, Flye WM. Renal transplants: can acute rejection and acute tubular necrosis be differentiated with MR imaging? *Radiology* 1991; **179**, 61–65.

13 Infection

The immune system that stands in the way of successful transplantation also stands guard against a wide variety of infective organisms. Just as splenectomy increases the risk of pneumococcal infection, so does the immune suppression of T cell function affect resistance to specific organisms. Cytomegalovirus, for example, produces minor ill health in normal individuals and yet may prove fatal in immunosuppressed transplant recipients. The capacity of transplant units to diagnose and treat these infections is central to successful results.

General principles

During a part of medical training one is carefully instructed on diligent and rigorous diagnosis of infections before consideration of treatment; yet in another the rewards of empirical treatment are paraded. The balance between broad spectrum treatment immediately and specific narrow spectrum treatment of a defined organism is most difficult in transplantation. On the one hand, patients who are heavily immunosuppressed in the early days and weeks after transplantation may decline very rapidly from presentation with a mild cough and fever, to death in a day. On the other hand, it is all too easy to find a patient committed to long term treatment with a noxious agent because of uncertainty over a diagnosis. In this chapter this issue is addressed in the context of each infection but an overriding set of principles can be espoused.

1. Early intervention is always warranted.
2. Early resort to invasive diagnostic modalities may be life saving.
3. Having collected all relevant pathological specimens, treat before the results are available.
4. Treat both the likely and possible pathogens until culture results are available.
5. Consider reduction of immunosuppression.
6. Save the patient first and the kidney second.
7. Recall that the temperature you are treating may be due to rejection, or to some other cause of pyrexia.

Prevention, prophylaxis and treatment

To a certain extent the infectious diseases that occur after renal transplantation are predictable. The balance of risk versus benefit of prophylactic agents alters as the frequency and predictability of specific infections rise. It is therefore in the field of transplantation that prophylaxis for a wide variety of pathogens becomes relevant. There are a number of preventive measures taken routinely by transplant units to reduce the chances of infection. These approaches must be seen as a continuum of effort designed to ameliorate the main cause of early morbidity and mortality after transplantation.

The hospital environment

The hospital can be seen as an efficient mechanism for concentrating a wide variety of pathogens and ensuring that those most resistant to treatment can flourish. To give one example of the impact of the hospital environment: hospital showers in infrequently used specialised transplant isolation rooms seem almost to have been designed for culture of legionella species. These organisms may then be aerosolised to infect the patient who has just recovered sufficiently to get out of bed for a shower. Regular attention to maintenance of hospital water systems with high temperature (60°C) hot water is thus essential.

Airborne pathogens may be excluded, to a certain extent, by positive pressure ventilation of hospital rooms and air filtration but few renal transplant units go to this extent.

Transmission of hospital pathogens on clothes and hands is, by contrast, a practical and proven approach to control. The simplest and probably the most effective measures are to reduce contacts to the minimum needed and ensure that all staff and visitors wash their hands. Some units achieve this end with elaborate gown and glove protocols, others by reminders to wash hands.

Food is a potential source of disease transmission even in the best run hospital catering establishments. Seafood and salad, raw or undercooked meats, some dairy products and illicit resort to 'take away' food are all worth elimination.

Animal pet contacts are usually best consigned to photographs, at least until patients return home.

The ultimate protection afforded bone marrow transplant recipients, with laminar air flow, sterile food and barrier nursing techniques, are no longer justifiable for routine renal transplantation. Isolation of a transplant recipient from patients with a known transmissible infection is of course, by contrast, mandatory.

The patient's endogenous flora

Cleaning the patient's skin before the transplant operation using bactericidal creams and soaps is a reasonable precaution. Taking swabs of potential pathogens may provide a vital clue to subsequent infection, especially if multiple resistant *Staphylococcus aureus* is found colonising skin or nose. Attempting to reduce bowel flora with non-absorbable antibiotics, on the other hand, is largely confined to bone marrow recipients and seldom used for renal transplantation. Candidiasis may disseminate from minor local colonisation to major systemic infection and for this reason some units make use of routine topical antifungal agents such as amphotericin lozenges.

The home environment

Home is generally a much safer place to be than hospital with respect to infectious organisms. There are, however, points to note for most patients. Pets and crowds of people are both risk factors for transmission of infection. Knowing that a patient's budgerigar has diarrhoea can be of critical importance and allow diagnosis and treatment of psittacosis, though avoidance would be wiser. The majority of infections caught from crowds are minor viral upper respiratory tract infections, which have little impact on the transplant patient except in the first weeks after transplantation. While it is usually impractical to avoid these infections, it is wise to ensure the patient does not deliberately make contact with those harbouring major infectious agents such as *Mycobacterium tuberculosis*.

Prophylactic measures

Vaccination, immune globulin, antiviral agents and antibiotics are the mainstays of prophylaxis in renal transplantation. Table 13.1 outlines some of the steps that may be used both before and after transplantation in order to reduce the subsequent likelihood of infection.

Wound infection threatens the physical environment of the graft and delays discharge from hospital. Good surgical practice is undoubtedly the best preventive measure, but a single dose of intravenous antibiotic with a broad spectrum, timed to peak at the time of surgery, is used by many to reduce the incidence of infection. Diabetic patients and those whose wounds are, for whatever reason, re-opened are most at risk of wound infection and thus require particular attention. The most important indication for prophylaxis is perhaps in patients suspected of harbouring tuberculosis, because of the impact of miliary tuberculosis resulting

Table 13.1 Prophylactic measures to prevent or ameliorate infection after transplantation

Vaccination – before transplantation	
Hepatitis B	Seroconversion difficult to achieve in uraemia
CMV	Not shown to influence CMV disease significantly
Hyperimmune immunoglobulin – after transplantation	
Varicella-zoster Ig	Contact treatment
CMV	Effect on actual disease uncertain
Prophylactic agents	
Acyclovir	Oral dose used for CMV and herpes virus prophylaxis
Co-trimoxazole	Urinary tract infection. *Pneumocystis pneumonii*
Amphotericin/nystatin	Oral candidiasis
Isoniazid (± other antituberculosis drugs)	Tuberculosis, if suspected on history or X-ray
Broad spectrum antibiotic e.g. Cefoxitin or other second generation cephalosporin	Wound infection

from immunosuppression. In parts of the world where tuberculosis is endemic it is routine to commence isoniazid at the same time as the immunosuppressive regimen.

Pneumocystis carinii pneumonia is a potentially lethal complication of immunosuppression which may effectively be eliminated as a risk by use of low doses of co-trimoxazole for the first six months. The further advantage of co-trimoxazole is that, when used in moderate doses (two tablets daily), it has been shown to reduce, by up to 50%, the incidence of urinary tract infection after transplantation.

Acyclovir and an antifungal agent such as amphotericin in the form of lozenges may be used to reduce risks from viral and candidal infection respectively. Practice varies considerably, reflecting conflicting views on efficacy and varying incidence of these infections in different units.

The timetable for infection

Not only are the organisms infecting transplant recipients predictable, but also the timing of each infection can be defined. It is thus possible to refine intuitive diagnosis (or best guess) before a definitive diagnosis can be achieved by culture, in order to treat the most likely organisms. In Table 13.2 the approximate timing of infections has been summarised. It is, however, important to realise that blind allegiance to such a schema will lead to errors and that this approach should only be used to guide the clinician.

Table 13.2 Timetable of infections after renal transplantation

0–30 days	30–180 days	>180 days
Wound infection	CMV – primary (at 42 days)	Community acquired:
Chest infection	– secondary (at 90 days)	viral infections e.g. influenza
Urinary tract infection	Herpes varicella/zoster	Bacterial infections
Intravenous line infection	Human herpes virus 6	e.g. pneumococcus
Septicaemia	Hepatitis C	Recurrent opportunistic disease
Herpes simplex virus	*Listeria monocytogenes*	inadequately treated and
CAPD catheter infection	*Legionella pneumophila*	not eradicated previously
Clostridium difficile	*Mycobacterium tuberculosis*	
enteritis	*Pneumocystis carinii*	
Salmonella	*Nocardia asteroides*	
	Candida, Aspergillus	
	Cryptococcus neoformans	
	Toxoplasma gondii	
	Strongyloides stercoralis (restricted)	
	Coccidioidomycosis (regional)	
	Histoplasmosis (incidence)	

Notes: Opportunistic infections which predominate in the period 30–180 days may be precipitated earlier than expected by use of a monoclonal or polyclonal antilymphocyte preparation. Approximate timings only.

Viral infection

Cytomegalovirus

Cytomegalovirus (CMV) is one of the most troublesome virus infections after transplantation, whether due to reactivation of latent disease or to transmission of virus either in the kidney or in transfused blood products. There are two ways of classifying CMV infection, either as 'primary' and 'secondary', or as 'infection' and 'disease'. Both approaches have their utility and their limitations.

CMV is usually acquired as a minor symptomatic infection before adulthood. The incidence of post-CMV infection is influenced by the patient's age, with between 50% and 80% of recipients having IgG antibody directed at CMV at the time of joining the transplant waiting list. In the immunocompetent host, CMV produces a minor illness that passes without notice as a mononucleosis-like syndrome. It, however, gains in clinical significance as a congenital infection, in transplant recipients, and in those with AIDS.

The four transplant situations are shown in Table 13.3 with 72% of recipients and 67% of donors having evidence of past infection with CMV. The 20% of patients who have never had CMV, receiving kidneys from positive donors, are at risk of developing 'primary' disease. All 72% of CMV positive recipients, irrespective of the donor status, are at risk of developing reactivation of the virus or 'secondary CMV'. In some elegant studies of CMV in transplant recipients it has been shown that the donor's strain of CMV may be different to the recipient's strain of CMV and thus it has been proved that re-infection with CMV, as opposed to reactivation of latent virus, is also possible in cases of 'secondary CMV'. While the kidney is the usual

Table 13.3   CMV disease and status of 113 transplant patients at Westmead Hospital

Recipient	Donor	Number	Percent	Potential Disease
Positive	Positive	53	47%	Secondary
Positive	Negative	28	25%	Secondary
Negative	Positive	23	20%	Primary
Negative	Negative	9	8%	–
		113	100%	

cause of virus transmission, whole blood, especially when fresh, may transmit CMV from a positive blood donor.

Immunosuppression with either polyclonal (ATG, ALG) or monoclonal (OKT3) antilymphocyte preparations is a further risk factor for development of CMV disease. In a high proportion of patients treated with these agents there is a 'viral syndrome' of pyrexia, malaise and myalgia which may or may not proceed to more significant disease and may or may not be associated with isolation of CMV. In the Westmead experience of OKT3 use, 50% of patients develop CMV infection after OKT3 therapy, with half of those showing signs of active CMV disease.

CMV infection   Infection with CMV implies that the patient has evidence of CMV such as; presence of IgM antibody specific to CMV; a fourfold rise in IgG antibody to CMV; or culture of CMV in blood or urine specimens. These findings may not be associated with apparent clinical symptoms or signs, even in an immunosuppressed host.

Primary CMV disease   In addition to evidence of CMV infection as described above, patients may develop the clinical picture of high swinging pyrexia with organ involvement either between four and eight weeks after the transplant, or five to ten days after ATG/OKT3. Primary CMV disease has been a serious clinical problem and a major cause of mortality in all transplant programmes. It is usual for bone marrow suppression to occur, needing a reduction in azathioprine. General constitutional symptoms include rigors, malaise and anorexia and it thus often proves difficult to distinguish CMV disease from superadded bacterial sepsis in this context. Serious complications of CMV disease include; hepatitis; pneumonitis; retinitis, leading to blindness; gastro-enteritis; encephalitis; and cardiomyopathy. Before introduction of effective therapy very few patients survived if they required ventilation for primary CMV. In this situation CMV is often detected in the urine or buffy coat and may be seen on kidney biopsy in which there may also be a CMV associated glomerulopathy.

Secondary CMV   Reactivation of latent virus by immunosuppression leads, classically, to a milder clinical disease with a peak incidence six to 12 weeks after transplantation of a patient positive for CMV specific IgG. Some cases of 'secondary' CMV, however, resemble 'primary' disease in both spectrum and severity of organ involvement, suggesting re-infection with a new strain of virus. Most cases have, on the other hand, a mild febrile illness lasting five to seven days, recovering without treatment.

Diagnosis   There are two ways of detecting CMV, either by direct isolation of the virus or by measuring a humoral response to the virus.

Serological methods can be used to detect evidence of past infection, in which case CMV specific IgG is elevated, or recent infection when IgM antibodies may be detected and the titre of specific IgG rises fourfold. Neither of these measures usually helps the clinician, except in retrospect, because it may take two to four weeks to achieve a diagnostic result, though detection of CMV specific IgM may be more rapid.

Histological appearance of renal biopsies or other tissue may be helpful since 'ground glass' intranuclear inclusion bodies can sometimes be seen. Specific immunoperoxidase stains for CMV have increased the specificity and sensitivity of histological diagnosis.

Culture of virus in human fibroblasts takes up to three weeks to produce a diagnostic cytopathic effect, though this may sometimes be seen sooner. Detection of CMV in fibroblasts has however been improved dramatically by addition of fluorescent anti-CMV antibody two to four days after initiating culture. Further advances in DNA technology have made it possible to identify specific CMV DNA or RNA in small copy numbers. It is possible that this new technology will prove too sensitive and detect CMV genome when no infection exists.

Treatment   Specific and effective treatment for CMV is now available using ganciclovir (Cymevene, Syntex), which is an acyclic nucleoside analogue dependent for activity on phosphorylation by cellular kinases. The cellular kinases are increased tenfold in virus infected cells thus maximising production of ganciclovir triphosphate in these cells. This triphosphate molecule is an analogue of deoxyguanosine triphosphate and is assembled into replicating viral DNA, causing instability and halting viral DNA synthesis. Dose schedules for ganciclovir are shown in Table 13.4, emphasising the fact that it is predominantly excreted in urine. Duration of treatment needs to be considered for each individual, but as a guideline

Table 13.4   Dosage of ganciclovir for treatment of CMV after renal transplantation

Serum creatinine (μmol/l)	Dose (mg/kg)	Dose interval (hours)	Dose/24 hrs (mg/kg)
< 124	15.0	12	10
125–225	2.5	12	5
226–398	2.5	24	2.5
> 398	1.25	24	1.25
Dialysis	1.25	Post dialysis	max 1.25

Precautions: Infuse intravenously over one hour in 100 ml normal saline or 5% dextrose
Monitor leucocyte count and platelet count
The solution has a high pH therefore avoid extravasation
Monitor liver enzymes, glucose
Drug interactions: zidovudine, probenecid, imipinem.

the minimum is two weeks and the usual period is three weeks. Reduction of immunosuppression, which used to be the mainstay of 'treatment' before the antiviral agents became available, has also to be judged on individual merits, but should not hazard graft rejection except in the sickest patients.

Alternative treatment with foscarnet has been used with success in patients with side effects from ganciclovir or where it is unavailable. It is a virostatic agent which inhibits herpes virus DNA polymerase, but it probably has a lower efficacy than ganciclovir and is associated with renal impairment, anaemia and hypercalcaemia.

Prevention   Primary disease may be prevented by providing CMV negative recipients with only CMV negative blood products and kidneys, or CMV positive blood filtered to remove leucocytes. This is not always a practical option and alternative strategies have been considered. Vaccination unfortunately yields poor dividends in uraemic patients, with only 20% seroconverting and a proportion of even those seroconverters getting CMV disease subsequently. Prophylactic administration of hyperimmune gammaglobulin either alone or in combination with acyclovir has been used, but also without dramatic impact. Prophylactic ganciclovir has been used with success in bone marrow transplantation but has not, because of the potential side effects, been used widely in prophylaxis for renal transplantation.

Epstein-Barr virus

The Epstein-Barr virus (EBV) may cause a primary infection similar to that seen in the non-immunosuppressed. The primary disease may occasionally progress to fatal infection, but is rare. Blood products, the allograft and conventional modes of transmission have all been implicated in transplant recipients. The major importance of this virus lies in reactivated disease in heavily immunosuppressed patients, especially after prolonged courses of antilymphocyte globulin. Uninhibited proliferation of B cells *in vivo* leads to oligoclonal immunoglobulin production and to lymphomatous spread involving such sites as the small bowel and central nervous system. The progress of these malignant lymphomas may be checked with a combination of acyclovir and withdrawal of immunosuppressants, but they are rapidly fatal if left uncontrolled.

Herpes simplex virus

Herpes simplex type I (HSV I), the predominant cause of oral and labial lesions, or HSV type II which mainly involves the anogenital areas, are common accompaniments of renal transplantation in the first few weeks and months. The incidence is higher in patients treated with larger doses of immunosuppressants and especially with antilymphocyte preparations. Cold sores create a nuisance for the patient and should be diagnosed and when progressing, treated aggressively, since spread to generalised oesophageal ulceration can occur. Diagnosis by immunofluorescent or electron microscopic techniques should be followed by treatment with acyclovir which may be given topically to external lesions, orally or if necessary intravenously. In the majority of patients five to seven days of treatment are sufficient.

Varicella-zoster

Reactivation of latent neurotropic virus to produce the familiar pattern of 'shingles' occurs ten times more frequently in transplant patients compared with the normal population. For a disease due to reactivation, in which new infection plays no part, it is surprising how frequent 'epidemics of shingles' are. The best clinicians will sometimes fail to predict the onset of the classical and unmistakable vesicular rash confined to a single dermatome in a proportion of patients presenting with prodromal pain which may be associated with malaise and pyrexia. Diagnosis in the laboratory from identifying the virus in vesicle fluid acts usually to confirm a secure clinical diagnosis. There are occasional patients in whom the rash is not classical, or in whom a rash never eventuates and in these patients changes in the antibody titre to herpes varicella-zoster may be the only indication. Trigeminal ophthalmic division zoster is a threat to the cornea and must be treated aggressively whether the patient is immunosuppressed or not. The danger for transplant recipients is of dissemination of virus, with all the consequences of severe chickenpox, while morbidity in the long term stems from prolonged postherpetic neuralgia.

The term chickenpox implies primary infection, while disseminated varicella-zoster implies secondary recrudescence. Both represent a serious and life threatening disease for transplant patients. The clinical picture may extend from minor illness with just a limited number of vesicular eruptions to severe illness with a confluent haemorrhagic and widespread vesicular and pustular rash, meningitis, encephalitis and pneumonitis. Diagnosis should not be delayed and involves examination of vesicular fluid. Treatment is now available to transform management of this condition. High dose intravenous acyclovir should start as soon as possible and continue for up to 14 days. Reduction of immunosuppression has to be judged for each individual and depends upon the severity of the illness as well as the degree and rapidity of response to acyclovir. New antiviral agents are being assessed and may offer alternatives to acyclovir in the future.

Prevention is an important strategy for transplant patients who have not previously had clinical chickenpox and who have no serological evidence of past varicella/zoster infection. Such patients should avoid exposure to virus and if exposed should be given prophylactic doses of zoster-hyperimmune gammaglobulin. The transplant unit has a responsibility to ensure that patients with zoster lesions are not managed in open wards or general transplant outpatient clinics, since it is highly contagious until all vesicles are crusted and dry.

Human herpes virus 6 (HHV 6)

A new addition to the herpes virus group is HHV 6 which has been implicated in a number of alternative minor clinical syndromes in healthy people and may be significant in transplant patients. There is clear evidence to show that some patients have a syndrome of fever and general malaise typical of mild CMV disease but in whom the titre of antibodies to HHV 6 instead of CMV rise. There is also evidence that HHV6 infection may act as a cofactor in determining the severity of CMV disease.

Human immunodeficiency virus (HIV)

There has been little controversy over the role of HIV in transplantation since a number of unfortunate individuals received transplants from HIV positive donors. This has happened in almost all countries where transplantation occurs and leads in most cases to death of the recipient in a matter of months as a result of opportunistic infection. Most centres believe that patients who are already HIV positive should not be transplanted for fear of precipitating the same fate when immunosuppression starts. Screening of donors for HIV is the single most important issue at present (Chapter 3) with an accurate personal and social history of donors being the only protection of recipients from a donor who has been infected within six to 12 weeks before donation, during which time antibody to HIV cannot be detected.

Hepatitis viruses

Hepatitis B and C may pose problems for patients on dialysis and have an impact both upon the decision to transplant and the patient's long term prognosis. Prevention of transmission of hepatitis B with the donor kidney has been mandatory clinical practice for many years using tests for Hepatitis B surface antigen. With the potential to screen donors for antibody to hepatitis C, many units are addressing their policy with respect to prevention of hepatitis C transmission.

Recent studies have shown that not only can hepatitis C be transmitted by an organ transplant, but that also the resulting infection may prove fatal. Up to 50% of recipients of organs from hepatitis C antibody positive donors have evidence of liver disease within six months.

While the case for screening out all donors with evidence of hepatitis B and C is undeniable, the situation is less clear with respect to the outcome of patients with either form of hepatitis before their transplant. Long term patient survival after transplantation is reduced in those positive for hepatitis Bs antigen at the time of transplantation. The question of whether or not transplantation should be offered is controversial. Evidence of active liver disease with raised transaminase levels and abnormal liver histology should probably be used as a relative contraindication to transplantation, though it is not clear whether the natural history is worse than with continued dialysis treatment.

Treatment of the infected patient with progressive liver disease is not generally effective. Reduction in the level of immunosuppression is usually advocated, but a measure such as interferon administration has yet to be shown to provide an advantage after transplantation, though it does increase the risk of acute rejection.

Other viruses

Adenovirus infection is common in the transplant population but does not carry significantly greater risk than is seen in the general population. Influenza A may be vaccinated against in transplant patients as part of general control of epidemic disease. There is no good evidence that vaccination after transplantation influences the outcome of the transplant or stimulates rejection and may thus be recommended to patients. Papillomaviruses cause common skin warts, which have an increased incidence in transplant recipients. More importantly, papillomaviruses are associated with both skin malignancy and genital tract tumours. Transmission of polyomaviruses by organ and corneal transplantation is well recorded, with progressive multifocal leuco-encephalopathy and relatively rapid death.

Bacterial infection

The majority of life threatening infections in the early period after transplantation may be attributed to bacterial infection. The patient is rendered susceptible to bacterial invasion not only by the immunosuppressive medication and previous uraemic state, but also by the surgical manipulations, urinary catheters and intravenous lines that characterise transplantation. A number of common pathogenic organisms provide significant threats to the transplant patient, in addition to which there is a group of opportunistic organisms that seldom cause clinical disease in the normal host. In Table 13.5 the common and less common infective bacteria have been listed. It is beyond the scope of this book to describe each in detail, especially where the clinical disease varies only in severity and speed of onset in immunosupppressed compared to normal people.

Table 13.5 Bacterial infection after renal transplantation

Common		Less common	
Gram positive			
Staphylococci		Mycobacteria	– tuberculosis
Streptococci			– atypical
Clostridium difficile			Anaerobic cocci
			Nocardia
Gram negative			
Enterobacteriacae	– Eschericia	*Neisseria meningitidis*	
	– Klebsiella	Salmonella	
	– Proteus	Anaerobic bacilli	
	– Serratia		
Pseudomonas			
Haemophilus			
Other			
Legionella		Listeria	
		Chlamydia psittacosis	
		Mycoplasma	

Gram positive bacteria

Staphylococci are the commonest cause of septicaemia, wound and intra-venous line infections in the immediate postoperative period. In hospitals where multiple antimicrobial resistance (MRSA) is common, patients are increasingly requiring vancomycin as a first choice antibiotic, on the presumption that MRSA is involved in a high proportion of cases. The problem of cross infection of patients with MRSA is a difficult and serious one. In units with a high incidence of MRSA, routine swabs of nose, axillae and groin reveal MRSA, which can usually only be cleared temporarily by local measures such as antiseptic baths and use of antiseptic nasal creams. Transmission is probably reduced most effectively by careful hand washing by staff and by nursing of those with frank infections, such as wound infections, in single rooms by staff who do not also nurse uninfected transplant recipients.

Clostridium difficile is a relatively frequent cause of diarrhoea in trans-plant patients, which may progress to bloody diarrhoea and frank pseudomembranous colitis if left untreated. Use of broad spectrum antibi-otics, which clear bowel flora and permit *Clostridium difficile* overgrowth, is a common risk factor. It is wise to assume that all antibiotic associated diarrhoea in transplant recipients may be due to *Clostridium difficile*, and to test for toxin in the stool.

Gram negative bacteria

Gram negative bacteria may be responsible for urinary tract infection, pneumonia, wound infection and septicaemia. Meningitis, cholangitis and osteomyelitis are less common but potentially devastating consequences of Gram negative infections. The implication of this wide spectrum of disease is that almost all systemic infections in transplant recipients require effective antibiotic cover of Gram negative organisms. The usual choice is a combination of an aminoglycoside together with ticarcillin or piperacillin, especially when pseudomonas is considered possible. The exact aminoglycoside choice should be determined with knowledge of the local hospital patterns of sensitivity and resistance, and doses based upon the level of renal function. An alternative strategy is to use a third generation cephalosporin, but this would not cover well for pseudomonas.

Legionella pneumophila

Legionnaires' disease in transplant recipients is common enough for consid-eration in the differential diagnosis of all chest infections. *Legionella pneumophila* lives in water and can survive at temperatures between 5 and 60°C. This is the range of temperature found in water cooling in air conditioning systems. Aerosolised, infected water may be inhaled leading to malaise, fever and general constitutional symptoms two to ten days later. This is followed by cough, haemoptysis, chest pain and dyspnoea. Systemic involvement may then lead to delirium, pericarditis, endocarditis, myocar-ditis, renal failure and gastro-intestinal disease. Hyponatraemia is character-istic and a useful clinical pointer to the diagnosis. Chest radiography may show patchy or lobar consolidation, progressing to widespread changes.

Diagnosis may be made by direct fluorescent antibody determination of sputum or lavage, but this is less sensitive than direct culture of broncho-alveolar lavage fluid. Detecting a rise in specific antibody titre is the more frequent method of confirming the diagnosis. Treatment requires erythro-mycin 1 gm every six hours by intravenous infusion with both rifampicin and doxycycline additional or alternative therapies in patients who fail to respond.

The untreated mortality is extremely high, but even in treated patients the elderly and heavily immunosuppressed may die despite treatment. Largely because of this possibility, erythromycin is almost mandatory therapy in a transplant recipient with pneumonia.

Nocardiosis

Nocardia asteroides are filamentous, Gram positive bacteria which normally live in soil, but may be inhaled or occasionally innoculated directly into skin wounds. It characteristically causes widespread nodular lesions in the lung or subcutaneous tissues. Haematogenous spread occurs in at least half of the infected patients, with cerebral abscess a frequent sequel. Diagnosis can only be achieved satisfactorily by direct microscopic examination and culture of specimens obtained from infected sites. Treatment is relatively straightforward with use of high dose sulphonamides, but amikacin, ampicillin and fucidic acid amongst others do have activity against nocardia. Response to treatment is generally slow and courses of therapy must therefore be prolonged.

Mycobacteria

Tuberculosis is a considerable problem in renal transplantation. Patients who have been infected many years previously and recovered, with or without treatment, may develop miliary tuberculosis as a result of immunosuppression. Prophylaxis for tuberculosis is thus the first consid-eration at the time of placing a patient on the waiting list. Patients who originate in areas of endemic tuberculosis need confirmation that they do not have active tuberculosis on chest radiography. Many units would then ensure prophylaxis is achieved with isoniazid for the first six months after transplantation.

Pulmonary tuberculosis may arise de novo after transplantation, with the usual presentation of cough, dyspnoea and haemoptysis. Dissemination is more frequent and more rapid in immunosuppressed patients compared to the general population. Treatment is essential but use of rifampicin is associ-ated with rapid metabolism of cyclosporin. It is possible to increase the dose of cyclosporin to counteract this effect, but a tenfold increase is needed. Consuming up to 50 ml of cyclosporin a day cannot be tolerated either by the patient or the financial authority, thus alternative immunosuppression or an alternative antituberculosis regimen must be used.

Atypical mycobacteria occur occasionally in immunosuppressed pat-ients causing skin lesions, lymphadenopathy or pulmonary disease. Cutaneous lesions associated with large shallow ulceration and sinuses from underlying lymphadenitis often respond poorly to chemotherapy. Surgical excision, even involving amputation of affected limbs, may be the only way to present dissemination and fatal disease.

Listeriosis

Listeria monocytogenes is a Gram positive bacillus which gains entry from the environment via the gastro-intestinal tract. Septicaemia and meningitis are the usual presentation in transplant recipients. Diagnosis by culture is time consuming and serum markers of infection often non-specific. Clinical suspicion and treatment of potential cases with ampicillin are the most effective ways of controlling the mortality.

Psittacosis

Chlamydia psittaci may infect poultry, pigeons and parrots causing a wasting disease with conjunctivitis, diarrhoea and a high mortality in these birds. In man the disease is characterised by a high fever, headache and pneumonia, with or without gastro-intestinal symptoms. Chest radiography shows extensive lobar or focal consolidation, which though characteristic is not helpful in discrimination of the diagnosis which requires serological confirmation. Tetracyclines, used for at least three weeks, are effective treatment, but preventation of contact in the early months after the transplant is important.

Mycoplasma

These organisms, without a cell wall sufficient to classify them as bacteria, may cause lower lobe pneumonia, general malaise, headache and an erythema multiforme-like rash. Formation of cold red cell agglutinins is a peculiarity useful in diagnosis, since both culture and specific antibody titres are slow and cumbersome methods. Since they lack cell walls they are indifferent to penicillins and other antibiotics that alter cell wall synthesis. Tetracyclines are effective, but erythromycin may be used in patients for whom tetracyclines are contraindicated.

Fungal infection

Fungal infection presents a spectrum of disease from the least troublesome to the most feared complication of immunosuppression. Oral candidiasis represents a minor irritant to the patient which can be treated early with a wide range of oral antifungal agents such as topical nystatin or amphotericin. Treatment of fungal septicaemia and disseminated infection requires long term use of toxic drugs.

Candidiasis

Candida albicans causes most of the fungal infection found in the early weeks after transplantation. It is a saprophytic yeast found as a normal flora commensal in the gastro-intestinal tract and vagina. After broad spectrum antibiotic treatment, however, the load of organisms rises. For this reason many units use prophylactic nystatin liquid or amphotericin lozenges orally to reduce the gastro-intestinal tract burden and lessen the risk of infection.

Culture of candida from the mouth, urinary catheter and vaginal swab may be sufficient to justify local or topical treatment in a transplant recipient. Oral candidiasis can spread to involve the oesophagus and cause significant symptoms, which can be diagnosed by endoscopy and treated with oral medication. Urinary culture may reveal candida in a catheterised patient or in a female patient with vaginal thrush for which antifungal pessaries are indicated. The difficulty in patients colonised with candida is when to consider systemic treatment. Clear evidence of candida in the blood stream or disseminated infection of lung and other organs (such as the eye) usually originates from intravenous devices and must be treated with intravenous amphotericin B or 5 flucytosine. Experience with fluconazole is accumulating to suggest it may be an alternative, and liposomal amphotericin, despite high cost, may have fewer side effects than conventional amphotericin. In patients where the evidence of candida in the blood is poor, one has to consider the costs and benefits of committing them to a prolonged course of amphotericin B.

Aspergillosis

The usual disease in transplanted patients infected with *Aspergillus fumigatus* starts with the lung as the portal of entry. Spread to the skin may allow rapid diagnosis by biopsy and culture of suspicious subcutaneous nodules. If unchecked, central nervous system involvement with brain abscesses and acute or subacute deterioration in function leads to death. Treatment with amphotericin B, sometimes even with the addition of rifampicin or 5 flucytosine, may fail to halt severe disease. A high index of suspicion, reduction of immunosuppression and early treatment are the most effective methods of preventing a fatal outcome.

Cryptococcosis

Subacute presentation of fever, headaches and meningitis symptoms are seen six or more months after the transplant. Subcutaneous nodules may precede meningitis by a considerable period as may involvement of the lung. Infection of the meninges may be evidenced by cerebrospinal fluid (CSF) with an increased lymphocyte count and detection of cryptococcal antigen. *Cryptococcus neoformans* may be cultured from the CSF but testing for antigen is less sensitive. Treatment traditionally involves intravenous amphotericin B but increasing use is being made of fluconazole for long term therapy.

Histoplasmosis

Histoplasmosis is endemic in parts of the United States but has occurred rarely in Africa, South and Central America, and Australia. Infection may lead to an acute pulmonary syndrome or may disseminate to bone marrow, liver and spleen.

Coccidioidomycosis

Like histoplasmosis the patients at risk for infection with *Coccidioides immitis* come from the American South West or semi-arid parts of South America. Pulmonary infection may be acute or chronic with dissemination to skin, joints, bone and meninges. Both amphotericin B and ketoconazole have been used successfully.

Parasitic infections

Transplanted patients may become infected with the same parasitic organisms that afflict normal people living in the same geographical area and sharing the same social and physical environment. There are, however, two organisms which provide transplant patients with a greater risk: *Toxoplasma gondii* and *Pneumocystis carinii*.

Toxoplasmosis

The cat is the host responsible for excretion of oocytes into soil. Infection of humans is usually due to ingestion in undercooked meats, with subsequent spread of sporozoites through the intestinal epithelium to all organs. Lymphadenopathy, hepatosplenomegaly, chorioretinitis, myocarditis and encephalitis are the hallmarks of progressive disease. Thirty percent of Australian patients, but up to 80% of French patients, will have antibody to *Toxoplasma*, depending upon their age. Diagnosis of primary disease rests on a rising antibody titre and presence of IgM as well as IgG. Identification of the organism in histological material may permit a diagnosis. Treatment involves pyrimethamine combined with sulphadiazine and is usually effective except in acute disease progressing rapidly to encephalitis.

Pneumocystis carinii

Infection with *Pneumocystis carinii* is extremely common judging by the high prevalence of antibody present in normal healthy children. Immunosuppression and AIDS have transformed this organism to a major cause of mortality. There are three forms of pneumocystis visible in infected lung; a 4–6 μm diameter cyst which may contain up to eight 1 μm sporozoites, and a larger trophozoite which is extracystic. The cyst wall is detected in specimens of lung tissue or broncho-alveolar lavage either by the Gomori stain with which it is black, or by toluidine-blue-0 with which it is lavender blue. Immunofluorescent tests are now available and may be more sensitive than morphological diagnosis.

Clinical presentation is characteristic and occurs three or four months after the transplant in a patient who has not been using co-trimoxazole prophylaxis. Severe tachypnoea, fever and a non-productive cough accompanied by low arterial P_AO_2 and a diffuse 'ground glass' infiltrate on the chest radiograph should alert the clinician. Diagnosis requires broncho-alveolar lavage performed without delay, as discussed in the section on Pneumonia below. Treatment is with co-trimoxazole given intravenously at a dose of 20 mg trimethoprim, 100 mg sulphamethoxazole per kg per day in divided doses. Fourteen days of treatment are usually recommended but improvement occurs in the first 48 hours. In patients allergic to sulphonamides, the alternative therapy is pentamidine, but hypoglycaemia, hypotension and nephrotoxicity relegate it to a second line. Prevention is infinitely preferable and can be achieved with a single tablet of co-trimoxazole daily for the first six months.

Differential diagnosis and management

There are a number of clinical presentations which characterise infections in immunosuppressed patients. The approach to diagnosis and management has to have full regard for both the differences in organisms and speed of progression in such patients, compared with normal immunocompetent patients. Approaches to the differential diagnosis of pneumonia, septicaemia, urinary tract, wound, neurological and gastro-intestinal infections are outlined here.

Pneumonia

The clinical presentation of a transplant patient with an acute onset of fever and cough, together with dyspnoea and constitutional symptoms such as malaise, anorexia and myalgia should precipitate a well organised and rapid assessment. Patients with minor symptoms and localised changes on a chest radiograph may progress to requiring ventilation within 24 hours without the correct treatment.

Historical aspects of importance include known recent or past contacts with viral upper-respiratory tract infections; tuberculosis; nosocomial contacts; pet birds with diarrhoea and weight loss (the avian presentation of psittacosis). Recent immunosuppressive drug therapy may include use of antilymphocyte preparations which favour development of CMV disease. Features of pneumonia seen in conventional patients should be considered, such as aspiration which may follow a period of coma, or a history of winter bronchitis. In specific patients, thought should be given to recurrence of systemic disease.

Symptoms may provide some assistance in diagnosis. Cough, fever and wheeze associated with dyspnoea are common to all diseases. Severe dyspnoea in an otherwise relatively asymptomatic patient should alert one to the possibility of pneumocystis pneumonia. Pleuritic chest pain, on the other hand, is more frequently associated with lobar pneumonia or pulmonary embolism. The speed of onset and severity of disease are helpful in assessing both the likely organisms and the appropriate diagnostic measures (e.g. fungal infections tend to be slower). Clinical examination may occasionally lead one to specific diagnoses. Lymphadenopathy may implicate tuberculosis or lymphoma, evidence of deep venous thrombosis suggests pulmonary embolus, but a high swinging pyrexia may accompany any infection.

The two investigations of paramount importance in initial assessment of the patient are an arterial blood gas measurement and a chest radiograph. The chest radiograph is mandatory and the resulting picture, combined with clinical progression, discriminates between likely diagnoses for a proportion of patients. A clear radiograph together with clinical symptoms and signs of bronchitis may be treated with antibiotics after sputum has been cultured. Amoxycillin, erythromycin and co-trimoxazole would each treat *Streptococcus pneumoniae* and *Haemophilus influenzae* which are common organisms in this situation. Careful follow-up is needed to ensure that recovery and not deterioration ensues. When the chest radiograph is not clear, however, it is essential to achieve a microbiological diagnosis using further investigations. A lobar distribution on the radiograph favours a conventional bacterial pneumonia or legionella, while a diffuse interstitial picture is more commonly due to *Pneumocystis carinii* or CMV infection. Between these extremes, a nodular or scattered infiltrate may represent infection with a variety of opportunistic organisms such as *Nocardia asteroides*, *Mycoplasma pneumoniae*, *Chlamydia psittaci*, or recrudescence of *Mycobacterium tuberculosis*. Culture of blood, urine, sputum, tracheal washings and (if tuberculosis is suspected) bone marrow aspirates, may all provide the diagnosis. There is no satisfactory replacement for broncho-alveolar lavage in diagnosis of a patient who is clearly infected and who has changes on the radiograph. Cytology of the fluid may demonstrate the characteristic appearance of pneumocystis or fungal hyphae. Immunofluorescence may help to detect CMV or legionella while culture will provide an uncontaminated result. Occasionally it may be necessary to resort to transbronchial or open lung biopsy to provide sufficient material for a diagnosis.

Treatment depends both upon the speed of progression and the state of the patient. A very sick patient clearly cannot await diagnosis by culture, while for others it may be more important to gain a diagnosis even if one or two days elapse before antibiotics are used. In a patient with significant symptoms, reduced gas exchange and infiltration on chest radiograph, broad spectrum antibiotics should be used immediately after the samples have been taken for culture. In the period from six weeks to six months after transplantation both erythromycin and co-trimoxazole should be used unless Legionnaires disease and pneumocystis pneumonia can be ruled out. These might be used in conjunction with a cephalosporin at least until a positive culture result becomes available. The major weakness of this combination is lack of activity against pseudomonas, for which an aminoglycoside and either ticarcillin or piperacillin are needed. These should be added if there is doubt about the bacteriological diagnosis in a patient with pneumonia and shock. CMV has already been discussed above. In the clinical setting of known generalised CMV disease, the onset of pulmonary disease may be attributed either to the virus or to superadded bacterial infection. The former may now be treated with ganciclovir, but the latter may be missed until there has been failure to respond to that treatment.

It is possible to monitor the state of the patient's pneumonia using clinical examination, arterial blood gas measurement, and chest radiograph. The latter takes the longest to resolve and usually remains abnormal long after the patient has been discharged.

Septicaemia

Presentation of a febrile and possibly neutropaenic transplant patient with tachycardia, hypotension and a poor cardiac output is a clinical emergency. A septicaemic patient may die very rapidly without treatment, which must therefore be instituted without delay. The major diagnostic issue to resolve, if possible, is the original site of infection. Urinary tract, surgical wound, chest and skin lesions are the usual sites of entry of organisms into the blood. Since the type of bacterial pathogen can be inferred from the site of entry an answer to this will guide the choice of antibiotic.

History, examination and investigation may be curtailed by an urgent need to resuscitate the patient. This should be conducted along conventional lines with intravenous fluids, inotropes and broad spectrum antibiotics. Corticosteroids will be needed to supplement the patient's routine medication, with hydrocortisone at a dose of 50–100 mg six hourly providing an acceptable short term boost. The use of high dose intravenous methylprednisolone in endotoxin shock is accepted in some units, but has a separate rationale to corticosteroid replacement at times of stress.

Diagnosis rests not only on history and examination but also on culture of blood, urine and any known collection of fluid such as a wound exudate. Patients with central intravenous lines should have them removed and replaced when septicaemia occurs. Culture of blood taken through the line and culture of samples taken from the tip of the removed catheter may provide a bacteriological diagnosis.

Neutropaenic patients are at greatest risk of Gram negative septicaemia, with *Escherichia coli*, *Klebsiella pneumoniae* and *Pseudomonas aeruginosa* the most common organisms which may be treated by an aminoglycoside and piperacillin or ticarcilllin. Fungal septicaemia and those caused by Gram positive organisms such as staphylococci are not covered well by this choice of therapy. The latter may be covered by flucloxacillin, though in hospitals with a high prevalence of multiple resistant organisms most clinicians would favour vancomycin. When fungal septicaemia is considered possible or probable it is reasonable to commence treatment while seeking definitive proof of invasion. Evidence for candida infection may come from knowledge of candida in the oropharynx, oesophagus, urine or vagina, but culture from the blood is needed to justify continuing a course of antifungal therapy.

Wound infection

Infection of the transplant wound may involve deep seated abscess formation around the vascular anastomoses or may involve only superficial skin edge cellulitis. Prophylaxis is of paramount importance, with surgical skill avoiding wound haematoma and lymphocele while antibiotics can provide appropriately timed cover for the procedure. Re-sutured wounds are at the greatest risk of becoming infected, emphasising the need for a single surgical procedure. Surgical drainage of the infected area is an essential prerequisite for resolution, while antibiotics have only a secondary role.

Urinary tract infection

Monitoring the urine of a renal transplant recipient may be achieved in three ways. Firstly, the clinical history detects changes in symptoms which may herald infection. Secondly, urinalysis dip sticks provide a relatively cheap and reliable method for demonstrating leucocytes in the urine or conversion of nitrites by bacteria. Thirdly, regular culture of midstream urine specimens in the first few weeks after transplantation allows subclinical infection to be detected. Judicious application of these three procedures ensures that infection of the urine can be treated by an appropriate antibiotic at the earliest opportunity. Particular care needs to be taken in patients with recurrent urinary tract infection before transplantation and in those with either native or transplant ureter reflux.

Meningo-encephalitis

There are two common settings for developing meningo-encephalitis. The first is related to treatment with OKT3 and presents as a headache with photophobia and neck stiffness, usually after the first one or two doses of OKT3 have been given. Cerebrospinal fluid (CSF) may reveal occasional lymphocytes but is otherwise normal in appearance and sterile on culture. Units differ on whether or not OKT3 is continued after diagnosis of aseptic meningitis, perhaps guided by the severity.

The second setting is a progressive onset of headache, fever and malaise during the first year of transplantation. Deterioration leads to neck stiffness,

Table 13.6 Treatment of some opportunistic infections in transplant recipients

Organism	Treatment of choice	Comments/alternatives
Legionella	Erythromycin	Rifampicin effective but interacts with cyclosporin
Pneumocystis	Co-trimoxazole	High doses required, pentamidine second choice
Mycoplasma	Erythromycin, tetracycline	
Clostridium difficile	Vancomycin, metronidazole	
Psittacosis	Tetracycline, erythromycin	
Toxoplasma	Pyrimethamine and sulphonamide	
Nocardia	Sulphonamides	
Listeria	Ampicillin, penicillin	
Herpes simplex	Acyclovir	Oral and topical
Herpes varicella/zoster	Acyclovir	High dose
Cytomegalovirus	Ganciclovir	Foscarnet

photophobia, neurological abnormalities and blunting of consciousness. Fundoscopy to exclude papilloedema, contrast computerised tomography of the brain, examination and culture of CSF, blood culture and assessment of serum for specific immunoglobulin titres and direct antigen usually provide the diagnosis. Electro-encephalography may help in diagnosis of herpes encephalitis but biopsy of the brain may be held as a last resort. *Cryptococcus neoformans, Listeria monocytogenes, Toxoplasma gondii*, mycobacteria, *Nocardia asteroides* and *Aspergillus fumigatus* would account for most of the opportunistic meningitides or cerebral abscesses. Viral encephalitis, in particular due to herpes simplex, may produce a devastating illness. Transplant recipients are of course not immune to the more usual infective organisms which may cause meningitis, such as pneumococcus and meningococcus.

Diarrhoea

Diarrhoea may plague transplant recipients as a result of infective and non-infective disease. While diagnosis and specific treatment are being determined, thought must be given to the consequences of possible malabsorption of immunosuppressive medications and dehydration which may accompany diarrhoea from any cause.

Clostridium difficile now only rarely progresses to cause perforating pseudomembranous colitis, but it is a common cause of diarrhoea after use of broad spectrum antibiotics. Diagnosis can be achieved both by culture and by identification of toxin. While treatment with either oral metronidazole 400 mg eight hourly or oral vancomycin 125 mg six hourly are effective, prevention is better. Cross contamination between patients and policies of broad spectrum antibiotic usage both need to be considered.

Amongst other pathogens which may lead to diarrhoea, campylobacter, salmonella species, cytomegalovirus and giardiasis account for most cases.

Oesophagitis

The infective causes of oesophagitis are usually either *Candida albicans* or herpes simplex. Clues to either may be found by examination of the mouth, with culture and microscopy of buccal scrapings providing diagnostic information. Fibre-optic endoscopy allows direct vision of the oesophagus and the chance of brushings of ulcers to diagnose their cause. The former may be treated with oral antifungal agents such as nystatin, ketoconazole (noting its interaction with cyclosporin) or amphotericin, while the latter will respond to acyclovir.

Further reading

1. Dunn DL, Mayoral JL, Gillingham KJ *et al*. Treatment of invasive cytomegalovirus disease in solid organ transplant patients with ganciclovir. *Transplantation* 1991; **51**, 98–106.
2. Weatherall DJ, Ledingham JGG, Warrell DA (eds). *Oxford Textbook of Medicine*. Oxford University Press, 1987.
3. Pereira BJG, Milford EL, Kirkman RL, Levey AS. Transmission of Hepatitis C Virus by organ transplantation. *New Engl J Med* 1991; **325**, 454–460.

14 Long term complications

The success of renal transplantation, in the early postoperative period, is almost entirely dependent upon two factors. The first is the patient's state of health before transplantation, in particular the states of the cardiovascular system and general nutrition. The second factor relates to the outcome of allograft rejection, its suppression by drugs and the consequences of immunosuppression. The long term outcome of renal transplantation is similarly influenced by the general state of health of the transplant recipient before transplantation, the effects of possible deteriorating transplant renal function, and the influences of long term immunosuppressive drug therapy. The greatest risks for mortality of transplant recipients, in the long term, emanate from cardiovascular disease, neoplasia, infections and, in some populations, liver disease. The risk of death is approximately 3 to 4% per annum beyond the first year after transplantation, depending upon the study and population examined. After six years, the renal transplant recipient is more likely to die with a functioning graft than lose the transplant because of chronic rejection.

In principle, the management of long term complications centres upon prevention, prophylaxis and treatment, though none of these strategies seems to have had much impact upon the rate at which patients die after transplantation. The slope of the survival curve for patients beyond the first year has improved little over the past 15 years. It would not, however, be entirely fair to presume that there has been no progress, since there are increasing numbers of patients now being transplanted that are both older and have more complex medical conditions than those transplanted 15 years ago. To achieve the same mortality in patients at greater risk of dying is thus some testament to the effort invested in reducing the risk of death after transplantation.

Cardiovascular disease

Incidence

The incidence of cardiovascular disease seen in a particular transplant centre is dependent, to a certain extent, upon the population of patients chosen for transplantation. Death from cardiovascular disease ranges from 30 to 50% of all late deaths after transplantation. The incidence of myocardial infarction after transplantation is lower if clear coronary arteries on a coronary angiogram are mandatory before joining the transplant waiting list. However, despite such screening methods, coronary artery disease remains a major problem, occurring with three times the frequency of an age and sex matched population without renal disease. The best predictive, but least sensitive, test for postoperative cardiovascular complications remains a clinical examination revealing pre-operative disease. Nevertheless in the presence of multiple risk factors for cardiovascular disease, it is almost certain that clinically relevant, if not clinically apparent, cardiovascular disease will exist.

Risk factors

The same risk factors that influence cardiovascular disease in the normal population affect the transplant population with presence of longstanding hypertension and diabetes the two most powerful predictors of risk. These are also the two factors that have the most significant impact on long term patient survival for all patients with endstage renal disease, irrespective of whether they are transplanted or not. Positive family history, male sex, obesity, smoking, age, hyperuricaemia and cholesterol levels (raised low density lipoproteins) have also been shown to increase the chance of cardiovascular disease in this group of patients.

A proportion of patients on dialysis develop significant large vessel calcification from prolonged hypercalcaemia and hyperphosphataemia. These patients are at considerable risk of cardiovascular events, but apart from parathyroidectomy, the most effective way of controlling this long term risk for the individual is renal transplantation.

After transplantation there has been debate over the role of cyclosporin in increasing the risk of coronary artery disease through its effects upon blood pressure, uric acid levels, glucose control and the vascular endothelium.

Prophylaxis and prevention

The patient carries the primary responsibility for changing three of the most important risk factors. Firstly, it is perhaps unreasonable to transplant a kidney into a patient who intends to shorten their life by smoking, if the kidney can be used for another patient without this risk factor. Surgeons vary in their opinion on whether to accept for transplantation a patient who smokes, but very few would omit an attempt to stop the patient smoking. Secondly, obesity is a remediable risk factor, and usually becomes more of a problem after transplantation as a consequence of corticosteroid therapy. The third factor that is partially controllable by the patient is the lipid level, which may respond to an alteration to diet. The new wave of lipid lowering agents, 'HMG CoA reductase inhibitors' (such as Simvastatin) and Gemfibrozil, have now given physicians the potential for inducing significant changes in cholesterol and triglyceride levels. It remains to be seen what role these drugs should be given, particularly after the initial unfavourable experiences in cardiac transplant recipients with Simvastatin. While the early studies suggested myositis was a particularly severe problem when Simvastatin was used in immunosuppressed patients, a number of units have cautiously combined these drugs successfully while monitoring blood muscle enzymes such as creatine phosphokinase. It is particularly relevant to resolve this issue since the adverse effect of cyclosporin on blood lipid levels has been well documented.

While one is clearly unable to influence sex, age and family history, these risk factors should be used to alert physicians to the possibility of significant asymptomatic disease. The control of hypertension, early recognition and optimum management of maturity onset diabetes mellitus and reduction of hyperuricaemia may, however, all need attention to reduce an individual's risk of cardiovascular events.

Management of hypertension

Control of hypertension is the single most time consuming medical problem in long term renal transplant recipients. While most patients with chronic renal failure before the need for dialysis have some degree of hypertension, only about one third of patients need antihypertensive drugs once they have commenced dialysis. Control of salt and water by dialysis is thus the most effective method of management. After transplantation, however, at least 80% of patients are hypertensive, with some becoming hypertensive for the first time. The reasons for this high incidence of hypertension are many (Table 14.1).

The original renal disease and associated hypertension may have lead, before the transplant, to structural arterial adaptations which result in persisting high blood pressure. The native kidneys, even without renal artery

Table 14.1   Causes of hypertension after renal transplantation

Allograft disease	– acute rejection
	– chronic rejection
	– recurrent disease
Native renal disease	
Renal allograft artery stenosis	
Drug induced	– Corticosteroids
	– Cyclosporin
	– FK506

stenosis, may contribute significant stimulation to the renin–angiotensin system and their radiological embolectomy or surgical nephrectomy may reduce the level of blood pressure. Both corticosteroids and cyclosporin are implicated in the genesis of hypertension in the transplant recipient. Dosage reduction is only partially successful as a strategy since there is evidence to suggest that the initial period of high dose corticosteroid predisposes to prolonged hypertension. It may be possible to reduce the dose of cyclosporin in some patients, but one does in the end have to provide satisfactory immunosuppression. Renal transplant artery stenosis is a cause of hypertension in a small proportion of patients.

Most studies of the transplant renal artery have shown that up to half of the patients with stenosis were not hypertensive. A new graft bruit, severe uncontrollable hypertension, or rapid decline in function following institution of an angiotensin converting enzyme inhibitor should alert the physician to the possibility of stenosis. Slow perfusion on DTPA scanning of the kidney, or a high peak velocity of main artery flow on Doppler ultrasound, may be used as initial or screening tests for renal artery stenosis (see Chapters 11 and 12).

Drug therapy for hypertension can be approached in much the same way as one would for patients before transplantation. There are, however, some issues which need to be considered in those with an allograft, usually because of drug interactions with cyclosporin. In the first six months after transplantation up to 80% of patients will require one or more drugs to reduce blood pressure. Each transplant unit will have its own preferences for drugs of first and second lines of therapy. Standardisation of the approach and being familiar with use of selected drugs is probably much more important than selection based upon nuances of theoretical but not practical relevance. Some of the drugs that may be used are discussed in Chapter 8.

Management of coronary artery disease

Management of ischaemic heart disease in a renal transplant recipient is no different to any other patient. The indications for therapy, investigation and surgery should not be adjusted in the stable long term transplant recipient. It is, however, worth taking special care with the state of hydration and diuresis during coronary angiography since the transplant is susceptible to contrast nephropathy and acute tubular necrosis.

Neoplasia

During the past ten years it has become apparent just how high the incidence of malignant disease is after renal transplantation. The data have been collected in three ways, each with its merits. Firstly, there have been single centre reports of neoplasias occurring with unusually high frequency, such as the lymphomas. Secondly, tumour registries have been established to record individual cases of malignant tumours reported to the registries, to document the range and relative frequencies of each type of cancer. Thirdly, national registries, of which the Australia and

New Zealand Data (ANZDATA) Registry is perhaps the best example, have collected tumour data on complete populations of patients.

The advantage of assessing complete transplant population data is that it allows calculation of relative risks of developing each tumour type (Table 14.2). Figure 14.1 shows ANZDATA Registry figures for the percentage of patients who have developed one or more histologically proven cancers from the time of transplantation. The intercept at which point all patients would have had at least one cancer is approximately 32 years. While this figure may seem a little academic, the fact that half of the Australian patients had had a cancer by about the 15 year point, and by that time more than 10% had cancer other than a skin cancer, gives cause for much concern. Australia is acting as a natural laboratory for studying the behaviour of such cancers because of the coincidence of fair skin and high levels of ultraviolet irradiation.

Skin cancer

In the normal population, skin cancers are usually basal cell carcinomas, almost always growing slowly and almost never associated with metastases. In the transplant population, by contrast, squamous cell are more frequent and are often aggressive and locally invasive (Figure 14.2). Up to 8% metastasise and half of those patients die within a year of metastasis. Unusual histology and rapid development of malignant disease from kerato-acanthoma and premalignant conditions such as Bowen's disease also characterise skin tumours in transplant recipients.

Sun exposure is the single largest risk factor, but even in Scotland, Sweden and Alaska the risk of skin cancer is such that patients should not feel secure without a widebrimmed hat, long sleeves and continued use of barrier skin creams. The back of the hands, face, ears and arms are the commonest sites for lesions. Continued vigilance by the patient is the most important preventive measure, since a rapidly growing cancer can intervene between routine follow-up clinics. Excision and histopathological

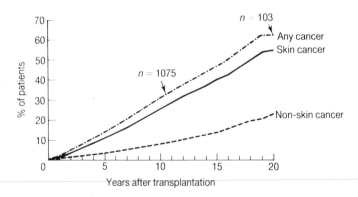

Figure 14.1   Cumulative incidence of patients who develop skin cancer, non-skin cancer or any cancer following transplantation. The number of patients followed to ten years (n = 1075) and 20 years (n = 103) are noted. Data supplied by the ANZDATA registry on all surviving Australian patients followed by six monthly questionnaire.

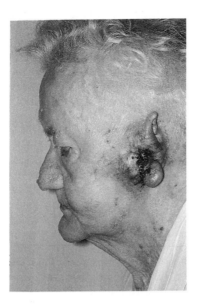

Figure 14.2 Advanced squamous cell carcinoma of the skin.

confirmation of any suspicious lesion should be recommended, since even well practised clinicians cannot distinguish malignant from non-malignant disease by eye.

There does come a time in some patients when the speed of growth and number of lesions preclude surgical excision and use of cryotherapy is the only practical solution. Reduction of immunosuppression has a clear role in patients with recurrent skin cancers but few are prepared to cease entirely and risk graft loss for what is usually a locally treatable disease. A number of centres have used Retinoates in an attempt to slow aggressive skin tumours, but convincing data to support this approach are not available.

Lymphomas

Lymphoproliferative disorders have been commonly reported after transplantation especially by transplant programmes which have used large doses of immunosuppression. It was thus from the use of prolonged antithymocyte globulin or combinations of high doses of potent immunosuppressants that lymphomas began to be reported. Cardiac and cardiopulmonary transplant recipients, perhaps because of their relatively heavy immunosuppressive protocols, have an overall incidence of 4 to 5% and appear to have the highest incidence of any transplant group. Recent data suggest that use of more than 75 mg of OKT3 in a patient is associated with a steep increase in the risk, to as high as 36%. Like neoplasias in the skin, the lymphomas in transplant recipients are different from those seen in the general population. Hodgkins lymphoma accounts for only about 2% of lymphoma after transplantation, compared with the 30% or more usually observed. Most are large cell non-Hodgkins lymphomas, and a disproportionate number occur in the central nervous system. About half are disseminated to multiple sites by the time of diagnosis.

These disorders have been grouped together under the term post-transplant lymphoproliferative disease (PTLD) and attention paid to observations on the degree of mono or polyclonality and the presence or absence of Epstein-Barr virus (EBV) infection. In general, polyclonal PTLD that can be shown to be associated with EBV infection has the better prognosis. By contrast monoclonal PTLD may be very aggressive and lead rapidly to the patient's death.

The therapeutic dilemma in patients with PTLD is profound. On the one hand, cessation of immunosuppression and graft nephrectomy with or without use of high doses of ganciclovir or acyclovir may lead to complete regression of the tumour. On the other, these may not be successful and result only in condemning a patient to dialysis in their last few months, as well as to death. The tendency is to choose a middle course with antiviral therapy and reduction of immunosuppression to a low level. It should be realised that each patient is different with respect to histology and site of the tumour, relationship to EBV infection and amount of immunosuppression used, such that no general guidelines can be relied upon to be applicable to each patient.

Long term cancer surveillance

It becomes increasingly important to monitor patients for tumour development as time passes after transplantation. The tumours which need monitoring for are documented in Table 14.2. Though these are not the only forms of cancer reported after transplantation, the list includes all of those where the relative risk for a transplant patient exceeds three (with the exception of female breast cancer which has been included for comparison with male breast cancer).

The mechanisms by which long term surveillance is achieved vary from centre to centre. The patient clearly has a major role in self examination and alerting the physician to potential problems. To achieve this, the patient needs to be taught about the importance of enlarging skin lesions, developing masses, changes in bowel habit and macroscopic haematuria. The role of the physician in surveillance is to take an accurate history; to ensure a thorough physical examination of the patient on a regular basis; monitor the urine for haematuria and abnormal urinary cytology; look at an annual chest X-ray; and investigate any significant symptom with speed and diligence (Table 14.3). In all female patients examination of the breasts (with mammography where available) together with a gynaecological examination and cervical smear are needed annually. Any patient with analgesic nephropathy as a cause of renal failure, particularly if haematuria is noted, will require an annual check cystoscopy and retrograde examination of the ureter to monitor for the possible development of urothelial tumours in the kidney, ureter or bladder. In those with known risk factors for development of liver disease or with known cirrhosis, a liver scan and ultrasound may be justified.

Musculoskeletal disease

Many patients complain of non-specific musculoskeletal symptoms: some are associated with return to a more physically active lifestyle, but others may be the early symptoms of significant disease with poor long term outcomes. It is important to attempt to distinguish these two situations clinically and without resort to multiple and repetitive investigations.

Avascular necrosis of bone

Avascular necrosis of the femoral head, leading to total hip replacement, was one of the hallmarks of a 'successful' renal transplant recipient during the late 1970s. The two prerequisites were survival long enough for the effects to progress far enough and use of high doses of oral and intravenous corticosteroids. In the high dose corticosteroid regimens of that era, the incidence was as high as 15%, while current low dose approaches are associated with a much lower incidence of 1% or less. The bones most often affected are the femoral head, tibial plateau (Figure 14.3), ankles, and scaphoid in order of frequency.

Pain is the usual presenting symptom, while diagnosis rests upon conventional X-ray which may reveal little in the early stages and bone scan (see Chapter 12) which may demonstrate an area of decreased uptake surrounded by an area of increased tracer uptake.

Analgesia, non-weight bearing crutches and reduced activity are the

Table 14.2 Relative risk of cancer after renal transplantation compared with an age and sex matched normal population. Data reproduced from the ANZDATA registry with permission

Site of cancer	Risk ratio (observed/expected)
Oesophagus	8.0
Small intestine	3.2
Liver	9.2
Pleura	8.5
Bone	7.5
Breast – female	1.1
Breast – male	4.6
Cervix – in situ	3.6
Cervix – invasive	4.5
Vulva and vagina	35.5
Bladder	4.9
Kidney	9.2
Ureter	198
Lymphoma – CNS	>1000
Thyroid	323
Diffuse non-Hodgkins Lymphoma	7.4
Multiple myeloma	4.3
Leukaemia	6.4
Kaposi's sarcoma	>1000
Melanoma	4.4

inadequate measures used for treatment. Total hip replacement or knee replacement provides, however, the best long term solution for patients with disabling disease of those joints.

Osteopaenia

Loss of bone mineral density (osteoporosis, or more accurately termed osteopaenia) is a problem well known to relate to the use of high dose and prolonged corticosteroid therapy. In transplant patients, osteopaenia is a more significant problem when high dose corticosteroids have been used than with current triple therapy regimens. Decreased osteoblastic activity but continued osteoclast bone resorption does, even with low dose prednisolone, lead to progressive reduction in bone density and to an increased risk of fractures of the hip and collapse of vertebrae.

Hyperparathyroidism

Patients present for renal transplantation with widely varying stages of

Table 14.3   Long term surveillance beyond the second year issues and investigations

Problem	Investigations	Interval between investigations if normal previously
Cardiovascular disease	Blood pressure	3 months
	Chest X-ray	12 months
	Electrocardiogram	12 months
	Fasting lipids	6–12 months
	Serum uric acid	6–12 months
New diabetes mellitus	Urinalysis	3 months
	Blood glucose (random)	6–12 months
Neoplasia		
General and skin	Full examination	6–12 months (+ patient)
General	Chest X–ray	12 months
Cervix	Cervical smear	12 months
Breast	Mammography	12 months
Urothelial	Urinalysis	3 months
Urothelial (analgesic nephropath)	Cystoscopy & retrograde pyelography of native kidneys	12 months
Ocular disease		
Cataracts	Ophthalmoscopy	3 months
Diabetic retinopathy	Ophthalmologist review	6–12 months
Liver disease	Liver function tests	3 months
	Liver scan and U/S	12 months
Haematological disease	Full blood count and film	3 months
Renal transplant disease	Creatinine, urea, electrolytes	3 months
	GFR estimate	12 months
	Urinalysis – protein	3 months
	– nitrites, leucocytes	3 months

* Note: This table is a general guideline for consideration and should not be taken as definitive advice on the investigation and management of individual patients.

hyperparathyroidism and renal osteodystrophy. In the long term, a successful renal transplant does, almost always, reverse this disease and return the parathyroid hormone secretion to normal levels. In the short and medium term, those with uncontrolled hyperparathyroidism may become seriously hypercalcaemic, while those who have undergone parathyroidectomy recently may risk dangerous hypocalcaemia. Hypercalcaemia may result from return to normal of both phosphate excretion and calcitriol production by the transplanted kidney. In many patients it is possible to observe the calcium level while the tertiary hyperparathyroidism involutes. In a minority, when the total corrected calcium remains above 3.0 mmol/l, parathyroidectomy may be needed. On the other hand hypocalcaemia is a short term problem needing treatment with intravenous calcium in the early postoperative period, and then oral calcium carbonate and calcitriol in the medium term.

Tendon and muscle disease

Corticosteroids are responsible both for muscle and tendon rupture and for generalised, usually proximal, myopathy. Rupture of the tendo-achilles requires primary suture repair if detected early, but is often seen days or weeks later, by which time suture is unsuccessful. Other muscles may be affected, but generally have a less debilitating effect on the patient than the

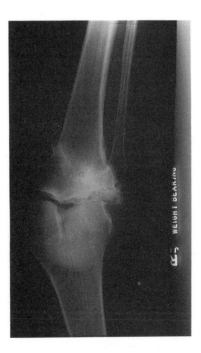

Figure 14.3 X-ray demonstrating collapse of the lateral tibial plateau in a patient with osteonecrosis.

achilles rupture. Proximal myopathy is a serious but fortunately uncommon effect of low dose corticosteroid therapy and can be diagnosed clinically, confirmed by muscle biopsy and treated by dose reduction.

Ocular disease

Cataracts

Posterior subcapsular cataracts are correlated with total doses of corticosteroids received by patients and occur in 30 to 40%. Lens extraction is needed in about half of those with symptomatic disease and occurs in up to 5% of transplant recipients. Most centres monitor patients' lens opacities regularly and attempt steroid reduction or withdrawal in those with developing lesions, though evidence that this is useful is hard to find.

Other ocular complications

The major long term ocular disease after renal transplantation other than cataract formation is diabetic retinopathy. Almost all patients with diabetic nephropathy have significant retinopathy, such that absence of retinopathy makes one seriously question the diagnosis of diabetic nephropathy. Correction of uraemia has little effect on progression of retinopathy and even when combined with a successful pancreas transplant, it is of vital importance to continue close and expert follow-up of retinal disease. Laser photocoagulation may be needed as either a prophylactic or therapeutic measure.

Hepatic disease

When the outcome of renal transplantation was poor, few patients survived with their grafts long enough to develop significant liver disease. It has since become apparent that chronic liver disease is a major long term risk in 10 to 15% of recipients of renal transplants, with the majority of chronic disease attributable to hepatitis B and C, while hepatotoxic drugs and cytomegalovirus may contribute in the short term. Both azathioprine and cyclosporin have been noted to be directly hepatotoxic and undoubtedly are in the experimental situation, but should not be blamed without careful exclusion of other much more likely causes.

Hepatitis B virus

Exclusion of all donors of blood and kidneys who are positive for hepatitis B surface antigen (HBsAg) and enquiry for behaviour which increases the risk of hepatitis B virus infection (HBV), has largely eliminated the risk of infections from this source. Routine vaccination of dialysis patients against HBV and isolation of those known to be positive for HBsAg have significantly reduced the number who present for transplantation with infection. Those patients who are persistent carriers of HBsAg at the time of transplantation have an increased risk of death from chronic liver failure in the long term. The fact that those patients paradoxically also have a higher early graft survival rate, presumably relating to increased suppression of the immune system in HBsAg positive patients, adds confusion to the literature. It is possible that presence of HBsAg and histological evidence of chronic active hepatitis at the time of transplantation both act as indicators of which HBsAg positive patients are at highest risk of progressive disease. Whether such patients die faster with a transplant or on dialysis is still a subject of debate.

Hepatitis C virus

Until the recent introduction of routine hepatitis C (HCV) antibody testing of donated blood and organs, the risk of transmission of non-A, non-B hepatitis was perhaps in the order of 1% in volunteer donor programmes. It has now been clearly shown that both blood and kidneys from patients with antibody to HCV can transmit the infection. In patients who are already infected with the virus, it is likely that liver disease will progress in a manner analagous to HBV but perhaps with a longer course. The number of patients infected with HCV is much larger than HBV, with estimates ranging from 1 to 15% of patients on dialysis; thus HCV is likely to become the most frequent cause of chronic liver disease in renal transplant recipients. While interferon has been shown to have an effect on the course of HCV infection it is also quite an effective agent for stimulating allograft rejection; thus the physician remains largely unable to change the prognosis for these patients.

Haematological disease

The effects of long term immunosuppression therapy on the bone marrow are occasionally dramatic and unpredictable, such as the sixfold increase in risk of leukaemia. Most, however, are slow to evolve, predictable, dose dependent and manageable by dose reduction of the offending drug.

Bone marrow failure

Azathioprine has a dose dependent effect upon all elements of bone marrow. It reliably produces a megaloblastic blood film which is not due to either folate or vitamin B12 deficiency. Isolated leucopaenia, anaemia and thrombocytopaenia can occur, though depression of the lymphocyte (and total leucocyte) count is by far the most common. During the first three months and at times of cytomegalovirus infection, leucopaenia is common. However, in the long term, azathioprine at doses of less than 2 mg/kg/day seldom cause significant problems.

Polycythaemia

Uncontrolled production of erythropoietin by the renal transplant is responsible for polycythaemia in up to 15% of patients. Until recently, this created a need for recurrent venesection when patients reached a threshold haematocrit (usually about 50%), at which their physicians became uncomfortable with the increased risk of thrombosis and thrombo-embolism. After a period of one to two years of repeated venesection, most patients regain feedback control of erythropoietin production and return to normal levels of haemoglobin. A number of centres have recently reported that a low dose of an angiotensin converting enzyme inhibitor will control polycythaemia within four to eight weeks. This is now the treatment of choice.

Gastro-intestinal disease

Mortality from gastro-intestinal disease has declined rapidly during the 1980s. There has also been a significant change in the pattern of disease such that peptic ulceration and diverticulitis, which were common, have now been replaced by such problems as the gastro-intestinal effects of CMV infection. The factors contributing to these changes have been dramatic reductions in doses of corticosteroids, increased use of antilymphocyte preparations, widespread use of effective anti-ulcer therapies and a better understanding of the pathological effects of *Clostridium difficile*. Nevertheless continued vigilance is needed to protect patients from untimely death.

Peptic ulceration

Presentation with a perforated or bleeding ulcer of the stomach or duodenum is now rare after transplantation. Prevention has centred around prophylaxis, effective diagnostic methods and early treatment. Most centres will investigate patients on the transplant waiting list by gastroscopy and duodenoscopy, if they have had significant symptoms of peptic ulceration. Following transplantation, epigastric pain and tender-ness need early investigation with biopsy and culture of specimens of oesophageal and gastric mucosa. The differential diagnosis usually lies between fungal infection (usually candida), viral infection (herpes simplex or cytomegalovirus) and peptic ulceration. Therapy thus needs to be directed at the specific diagnosis: nystatin or amphotericin orally, acyclovir or ganciclovir systemically, or oral antacid with an H_2 receptor antagonist respectively. Surgery has almost no place in the management of this complication.

Pancreatitis

Between half and 6% of renal transplant patients will develop pancreatitis and 70% of those will die as a result. While corticosteroids and azathioprine have received blame both in transplant recipients and outside of that context, cyclosporin has also been considered a culprit. The more usual associates of acute pancreatitis, such as cholelithiasis and high alcohol intake, do not provide adequate explanation for most cases occurring after renal transplantation. Management centres upon early recognition, parenteral nutrition, fasting and possible surgical removal of necrotic pancreas.

Colonic disease

The commonest colonic problem after transplantation was pseudomembranous colitis leading to diarrhoea, haemorrhage, perforation of the bowel and death. Greater understanding of the roles of broad spectrum antibiotics, *Clostridium difficile* and treatment with oral vancomycin or metronidazole has now consigned this disease to a transient diarrhoeal illness. Diverticular disease leading to abscess formation, perforation and peritonitis does remain a significant problem with a higher than expected mortality in immunosuppressed patients. Early and aggressive intervention is needed in this situation.

Diabetes mellitus

Up to 10% of patients who were not diabetic before a renal transplant subsequently develop glucose intolerance. The factors which contribute to this are pre-existing mild but unrecognised glucose intolerance, corticosteroid induced peripheral insulin resistance, cyclosporin pancreatic islet cell toxicity, and obesity. In a minority of patients, chronic pancreatitis may contribute, while a family history of diabetes and black or Hispanic ethnic origin are all more common in those who develop diabetes than those who do not. Almost all patients who become diabetic do so within a year of transplantation and it is thus relevant to monitor urine for glycosuria and random blood glucose. Glycosuria is quite common and often does not imply glucose intolerance but instead implies tubular defects in glucose handling which may be attributed to cyclosporin in some patients. Management of diabetes involves diet and weight reduction in all patients, oral hypoglycaemic agents in half and insulin in the other half.

Native renal disease

There have been many phases in the approach to bilateral native nephrectomy in transplant recipients. Initially it was common practice to remove the kidneys in any patient in whom they may provide a source of sepsis after transplantation. The dire long term effect of blood transfusion dependence in most anephric dialysis patients reversed this policy. It has again become prudent to consider a more liberal approach to nephrectomy where erythropoietin is available for use by the patient. In practice two causes of renal failure contribute most of the problems after transplantation: polycystic kidney disease and reflux nephropathy.

Polycystic kidney disease

Some patients will have come to transplantation only after unilateral or

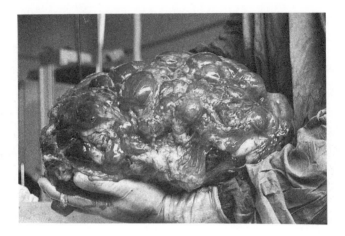

Figure 14.4 Polycystic kidney removed to ensure adequate space in the abdominal cavity for a renal transplant.

bilateral nephrectomy. The indications for nephrectomy include recurrent infection or haemorrhage into renal cysts, but usually centre on physical size and the potential for interference with renal transplant surgery (Figure 14.4). A small proportion of polycystic kidneys cause problems in the long term as a result of infection or haemorrhage, either of which may provide a diagnostic problem to distinguish native from transplant kidney disease. Nephrectomy may be needed for recurrent episodes of a severe nature.

Pyelonephritis

Recurrent urinary tract infection as a result of urinary reflux into the native ureters and kidneys may re-emerge after transplantation. Patients who have become oligo-anuric on dialysis may not have suffered urinary tract infections for many years. Establishing both urinary flow and immune suppression is sufficient to re-introduce urinary infection and both native and transplant kidney pyelonephritis. Bilateral nephro-ureterectomy may provide the only long term solution to this problem if urinary prophylaxis with simple antibiotic treatments fails to control infection (Figure 14.5).

Pregnancy

The return of fertility to both male and female patients is for some the main impetus to seek transplantation. While male transplant recipients have only to consider their own long term prognosis in relation to the needs of a family, the implications for females are greater. Pregnancy is associated with a small but real risk to the transplant, with acute rejection episodes occurring in about 10% and some impairment in function in up to 15%. The pregnancies are usually complicated, with a 15% incidence of first trimester abortion; intra-uterine growth retardation and premature delivery

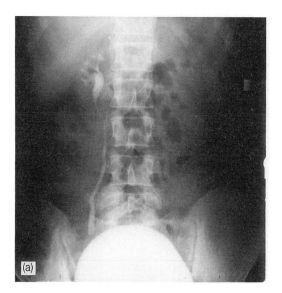 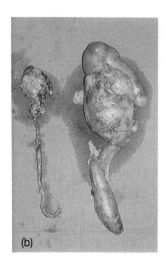

Figure 14.5 Micturating cystogram (a) in a patient with recurrent urinary tract infection, together with the bilateral nephrectomy specimen from this patient (b), demonstrating the obstructed left ureter with gross dilatation together with the shrunken remnants of the right kidney.

in half; and a reported increase of the risk of cleft palate in the child of a mother using azathioprine. Cyclosporin has a reversible effect on the child's renal function but angiotensin converting enzyme inhibitors must be stopped. Most transplant units would actively discourage pregnancy in the first year after transplantation, but advise those who have stable renal function after one year of the risks to themselves and chances of a successful outcome. The consequence of this advice is a need to provide effective access to contraception. Most dialysis patients rely on the effectiveness of uraemia as a contraceptive and many are unaware how quickly this may reverse, emphasising the need to address the subject with all those of childbearing age.

Further reading

1. Sheil AGR. Complications of immunosuppression in renal allograft recipient, malignancy. *Clin Transplant* 1991; **5**, 573–579.
2. Alfrey EJ, Friedman AL, Grossman RA *et al*. A recent decrease in the time to development of monomorphous and polymorphous post transplant lymphoproliferative disorder. *Transplantation* 1992; **54**, 250–253.
3. Julian BA, Laskow DA, Dubovsky J, Dubovsky EV, Curtis JJ, Quarles LD. Rapid loss of vertebral mineral density after renal transplantation. *N Engl J Med* 1991; **325**, 544–550.
4. Periera BJG, Milford EL, Kirkman RL, Levey AS. Transmission of Hepatitis C virus by organ transplantation. *N Engl J Med* 1991; **325**, 454–460.
5. Muirhead N, Sabharwal AR, Reider MJ, Lazarovits AI, Hollomby DJ. The outcome of pregnancy following renal transplantation – the experience of a single center. *Transplantation* 1992; **54**, 429–432.

15 Paediatric renal transplantation

The devastating effect of chronic renal failure on the mental and physical development of children offers unique and challenging management problems. Although less than perfect, successful renal transplantation is considered the best available treatment option for children of all ages, with the alternative of dialysis often made difficult by access problems, growth retardation and renal osteodystrophy. The greatest challenges are in the child less than five years of age, who is most likely to have a congenital urinary tract abnormality, and be below the third percentile for weight and height. It is in this age group that the effects of uraemia on development are greatest and results of transplantation have been consistently inferior. However, even in this younger age group, several large centres have recently demonstrated the advantage of an integrated approach to correction of medical and surgical problems associated with childhood endstage renal failure and early transplantation.

One year cadaver graft survival in excess of 80%, and even better results with living related donors, should now be expected. Nevertheless, the very good graft function needed to provide maximal rehabilitation is not easily achieved because of the nephrotoxic effect of cyclosporin, the need to minimise steroid immunosuppression and the high incidence of immunosuppression non-compliance. This chapter serves to highlight the differences between paediatric and adult renal transplantation, particularly for the young child recipient, and to demonstrate that the practice of paediatric renal transplantation is much more than a scaled down version of adult transplantation.

Primary renal disease

The incidence of childhood endstage renal disease presenting for treatment is of the order of two to five patients per million population (pmp) per year, and is relatively low compared to that in adults (up to 100 pmp). The distribution of primary renal disease varies markedly from that of adults with approximately 60% of children presenting with congenital or inherited disorders, and 40% with an acquired renal lesion (Table 15.1). Focal sclerosing glomerulonephritis is the commonest form of glomerulonephritis leading to endstage renal disease in childhood. The incidence of systemic lupus erythematosus has decreased as a result of aggressive immunosuppressive therapy, but there is no evidence that treatment has influenced other childhood acquired glomerular lesions. Comparison between transplant centres or registries is made difficult because of variability of upper age limit used for data collection and geographic locality. For example, congenital nephropathy and haemolytic uraemic syndrome have a high incidence in some European countries, whites have a high incidence of renal tract abnormality, and blacks a high incidence of glomerulonephritis. Nevertheless, the young child is more likely to have a congenital or inherited disorder and the older child, over the age of ten years, is both more likely to have renal failure (Figure 15.1) and more likely to have a glomerulonephritis. Identification of the aetiology of endstage renal disease is important, particularly when evaluating potential live related donors, because of the strong familial incidence of ureteric reflux and as in the case of focal sclerosing glomerulonephritis, a high incidence of recurrent disease.

Table 15.1 Distribution of causes primary renal disease (percentage incidence) in 221 children aged between 0 and 14 years in Australia from 1981 to 1990

Congenital and hereditary	63%
* Reflux nephropathy	19
* Hypoplasia and dysplasia	13
* Medullary cystic disease	9
* Posterior urethral valves	6
* Cystinosis	5
* Other (known)	9
Acquired	37%
* Glomerulonephritis	24
* Haemolytic uraemic sndrome	4
* Trauma and cortical necrosis	2
* Other (known)	7

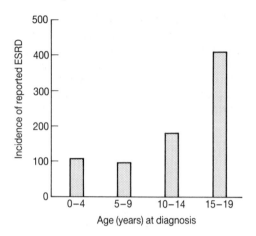

Figure 15.1   Age at the time of presentation for replacement renal therapy of paediatric USA Medicare patients (USRDS Annual Data Report 1990).

Management approach

To maximise childhood growth and development, early recognition of impaired renal function, before replacement renal therapy becomes necessary, is essential. Although the precise mechanism of uraemia associated growth retardation is unknown, it can be minimised by introduction of various therapeutic manoeuvres. Physical development can be assisted by nutritional support, correction of metabolic acidosis and an adequate intake of salt and water. Night time supplementation by nasogastric or gastrostomy feeding is often necessary. Bone disease can generally be controlled by dietary manipulation, use of phosphate binders (calcium carbonate) and vitamin D. Anaemia can be corrected with recombinant human erythropoietin which has the added advantage of limiting the need for multiple transfusions and thereby reducing the risk of sensitisation of the potential renal transplant recipient. Physical growth can be assisted by the use or recombinant growth hormone.

At the time of presentation for renal replacement therapy, many children will have already undergone previous surgical procedures to correct major urinary tract abnormalities. Infection and inadequate bladder capacity in such instances is common and should be assessed and definitively resolved before transplantation. The infected small bladder can sometimes be rehabilitated by regular distention and washout with antiseptic solutions but its adequacy should always be assessed by cysto-urethrography and urodynamic studies before contemplating transplantation. Attempts to reconstruct the urinary tract after transplantation are made difficult by the effects of immunosuppression on infection and healing, particularly when the transplant ureter is surrounded by scar tissue from previous surgery.

The considerable advantage of maintaining native bladder function after transplantation, and appreciation of the difficulties associated with reconstructive urinary tract surgery after the time of transplantation, may make difficult the decision to provide urinary diversion by construction of an ileal conduit beforehand. In most instances, however, the need for an ileal conduit is obvious and construction is undertaken at least eight weeks prior to renal transplantation. In such instances, the donor kidney is placed on the contralateral side from the ileal conduit and in an extra peritoneal position if possible. Through a separate midline incision, the ureter is brought into the peritoneal cavity, spatulated and anastomosed in a slightly upwards direction to the proximal end of the ileal conduit (Figure 15.2). The anastomosis is protected by a temporary catheter passed down the conduit, through the ureter and into the renal pelvis of the transplanted kidney.

Dialysis

The need for aggressive nutritional support of children with progressive renal disease is often the major indication for introduction of early dialysis in childhood. Whereas haemodialysis is technically feasible, even in very young children, difficulty is encountered frequently with creation of suitable vascular access. Thrombosis of synthetic vascular access devices is common. Hence, in most instances, continuous ambulatory or continuous cycling peritoneal dialysis is favoured as it provides continuous biochemical control, allows greater volume nutritional supplementation and can be performed outside the hospital environment. The disadvantage is the time commitment on the part of the child's parents. Furthermore, previous abdominal reconstructive surgery may make the peritoneum unsuitable for dialysis, and social circumstances may prevent adequate training for home peritoneal dialysis.

Recipient age as a criterion for transplantation

Whereas transplantation clearly provides the optimal treatment for the child aged greater than five years, management decisions for those younger are

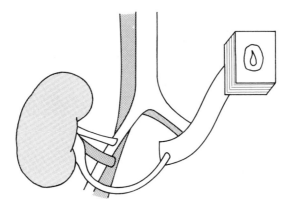

Figure 15.2 Renal transplant urinary diversion provided by anastomosis of the transplanted ureter to an ileal conduit which is fashioned at least eight weeks before renal transplantation.

more contentious. Data collected by transplant registries from children under the age of five years, particularly in those aged less than one year, demonstrate results that are less favourable than those in the older child. The less satisfactory transplant outcome in the younger age group has been attributed to technical problems and reputed enhancement of the immune response of the young child. Nevertheless, experience of the University of Minnesota transplant group would suggest that all children, irrespective of age, should receive a renal transplant, with preference in the younger age group being given to the use of living related donor kidneys. Their results, with minimal differential between age groups, offer a standard by which registry and individual transplant centre results should be compared. If the transplantation results of the young child are not comparable, transplantation may reasonably be deferred, provided the child is thriving and developing on peritoneal dialysis, until they are more than five years of age or weigh greater than 20 kg. It is, however, most unusual for children in this age group to thrive and gross motor delay is common.

For the very young child, therefore, transplantation should always be undertaken, if possible, once the child has reached endstage renal failure. This recommendation is supported by the USRDS Annual Data Report of 1990 which demonstrated that 49% of 132 children diagnosed with renal failure under the age of five had a functioning renal transplant two years later, 20% had died and the remainder were on dialysis. Peritoneal dialysis patients outnumbered haemodialysis patients by three to one. In the same age group, the two year survival for patients receiving living related transplants was 97%, and 89% for those receiving a cadaver renal transplant. On dialysis, the two year survival was reduced to 75%.

Donor sources

Transplantation of an appropriately sized cadaver kidney into a young child would seem the most practical option to overcome the technical and physiological challenges provided by placement of an adult kidney into a paediatric recipient. Unfortunately for the paediatric recipient, brain death in the paediatric age group, usually a result of trauma, is a relatively uncommon event. Anencephalic children are often unsatisfactory for use as renal donors because of difficulty in establishing a diagnosis of brain death, associated urinary tract malformations and frequent primary graft non-function. Furthermore, transplantation of kidneys from other very young donors is less successful because of the increased incidence of vascular thrombosis. This experience has prompted many centres, but not all, to set a minimum acceptable donor age of somewhere between one and five years. Sibling donors are not a practical option unless they have reached the legal age of consent.

Use of adult living related or cadaver kidneys in children weighing greater than 20 kg is usually without problem, with the lowest acceptable recipient weight being about 7 kg. The greater availability of adult donor kidneys provides the opportunity for earlier transplantation and better tissue antigen matching. In this respect, living related donor kidneys, usually from one haplotype matched parents, provide graft survival which is in the order of

10% better than that of cadaver donors. Furthermore, living related donors have the major advantage of permitting elective transplantation, and more importantly, it is becoming increasingly apparent from transplant registry data that long term graft survival in excess of 20 years is more likely to be achieved with a living related kidney.

Surgical procedure

When the donor kidney is of an anatomically appropriate size for the recipient, the technique of transplantation in the child is the same as for the adult. The more usual situation confronting the transplant surgeon, however, is the need to transplant an adult sized donor kidney. For children weighing greater than 20 kg, the extraperitoneal surgical approach using a conventional incision is usually feasible, with the renal artery and vein anastomosed to the lateral side of the recipient common iliac vessels. Transplantation of an adult donor kidney into a recipient weighing less than 20 kg is more difficult, with size disparity producing technical and physiological problems. It is in this group of recipients that graft survival rates are variable and probably reflect centre experience and commitment. The very best results are comparable to adult graft survival.

Planning the operation

The bulk of an adult sized donor kidney often precludes its placement in the pelvis of a young child, and for this reason, intraperitoneal positioning of the kidney through a midline abdominal incision is frequently necessary. Of principal concern is the need to avoid kinking or compression of the transplant vessels, which can most easily be achieved by placing the kidney in the lateral retroperitoneal space either behind the right or left colon depending on the side of the donor kidney. More appropriate donor vessel length is achieved if the donor kidney is transplanted into the ipsilateral side of the recipient, with emphasis on anastomosing the donor renal vein and artery to the lateral side of the inferior vena cava and infrarenal abdominal aorta respectively. Positioning may be facilitated by native nephrectomy on that side, and bilateral native nephrectomy may be indicated in the presence of chronically infected kidneys or uncontrolled nephrotic syndrome.

Vascular anastomoses

With a right sided kidney, many surgeons prefer to mobilise the inferior vena cava in the young child to allow the donor renal artery to pass posteriorly (Figure 15.3). On the left side, the renal vein is passed anterior to the abdominal aorta. A continuous prolene suture can be used for anastomosis of the renal artery if the aortic patch is present; however if absent, at least half the arterial anastomosis should be performed using interrupted sutures to allow for subsequent growth of the child.

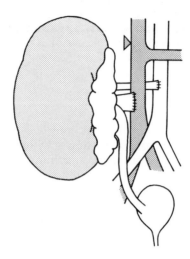

Figure 15.3 Illustration of surgical technique for transplantation of a right sided adult donor kidney into a young child following right native nephrectomy. The vascular anastomoses are to the abdominal aorta and vena cava.

Physiological stress

Revascularisation of a large cold adult donor kidney provides a major physiological stress for the young child recipient and requires careful intra-operative fluid management to compensate for the required increase in cardiac output. Before removing the vascular clamps, 250 ml of fresh whole blood or colloid should be transfused. Warming the donor kidney prior to transplantation is impractical because of the tubular damage produced by warm ischaemia. The combined vascular anastomosis time in children is usually in the order of 45 minutes and further warm ischaemia would be detrimental to the chances of primary renal function. Hence, all other appropriate measures should be taken to reduce the risk of hypothermia, including the use of a warm operating theatre, placement of the patient on a heating blanket, warming of all intravenous fluids and minimising exposure of bowel to operating room air with the use of warm moistened packs. The recipient temperature should be monitored by a rectal temperature probe. To assist diuresis and early renal function, intravenous bolus doses of mannitol 0.25 gm/kg and frusemide 1.0 mg/kg can be given at the time of revascularisation.

Other surgical considerations

The ureteroneocystostomy technique is the same as that used in the adult. The maximum size for an indwelling urethral catheter is usually between 8 and 14 gauge, thus presenting a problem for removal of a bladder clot after transplantation. The preferred option in a young child may therefore be the placement of a larger suprapubic catheter at the time of transplantation. Although acute appendicitis is not more common in paediatric transplant recipients, some surgeons prefer to remove the appendix at the completion of the transplant procedure because of the subsequent difficulty that may be experienced in diagnosing right sided abdominal pain. Prophylactic appendicectomy however, provides an unnecessary risk of contamination of the abdominal cavity in the presence of large doses of immunosuppressive agents.

Immunosuppression

Evaluation of immunosuppression protocols in paediatric renal transplantation is made difficult by the paucity of randomised controlled trials, and probably reflects the inability to obtain sufficient sample sizes to assess efficacy of treatment protocols. Published data from large paediatric transplant centres usually cover large time periods, include many variables and rely on historical data for comparison. Nevertheless, the consensus approach of most centres is to use an integrated combination immunosuppressive therapy protocol that allows flexiblilty to maximise early immunosuppression and to tailor dosages to meet individual requirements. For both cadaver and living related transplantation, the choice is between triple therapy of cyclosporin, azathioprine and prednisone, or sequential therapy using an antilymphocyte preparation to avoid the use of cyclosporin in first week of transplantation.

Children require not only a functioning graft if they are to maximise their growth and development, but also excellent graft function. The dilemma facing paediatric transplant clinicians is to prevent rejection and yet strive to maintain low doses of both cyclosporin to reduce nephrotoxicity and improve GFR, and corticosteroids to minimise the associated deleterious effect on skeletal growth. Suggested immunosuppressive protocols are outlined in Table 15.2. Many variations to these suggestions exist, including routine use of antilymphocyte preparations for ten to 14 days after transplantation, either with or without very low doses of cyclosporin, and followed by more appropriate doses in order to obtain recommended therapeutic blood levels.

Cyclosporin

Children metabolise cyclosporin more rapidly than adults, with those less than ten years of age having up to fourfold greater clearance by the cytochrome P450 oxidase system. For this reason, the dose per kilogram body weight is comparatively higher than for adults and must be given either in two or three divided doses per day. Although the cyclosporin dose can be reduced in a stepwise fashion in the early weeks after transplantation, therapeutic monitoring of blood levels provides for consistency and safety of cyclosporin use, particularly in the young child where liver dysfunction due to hepatic congestion or intrinsic hepatobiliary disease may reduce the rate of elimination of cyclosporin.

Suggested 12 hour trough whole blood levels of parent molecule cyclosporin are between 100 and 200 ng/ml in the first three months after transplantation, and reducing to between 50 and 100 ng/ml thereafter.

Whereas cyclosporin has the advantage of reducing the need for steroid medication, the nephrotoxic effect of higher doses will reduce the quality of graft function, a feature that is less important in adults but vital to a child's growth and development. Of further concern are the long term effects of associated hyperlipidaemia and hypercholesterolaemia.

Corticosteroids

Corticosteroids dosage is perhaps the most important variable affecting skeletal growth after transplantation. Initial low doses and even lower maintenance doses have been a major factor in almost eliminating the incidence of childhood avascular osteonecrosis. Furthermore, it has been demonstrated by the UCLA group that optimal growth requires the daily prednisone dose to be less than 0.24 mg/kg. Use of alternate day steroid therapy has been reported to be associated with a decreased incidence of hypertension, lower cholesterol levels, improved cosmetic appearance and improved growth, but unfortunately, there is a tendency for recipients with long term functioning grafts not to comply with this dosage regimen. An alternative is careful withdrawal of steroid therapy in the first three to six months after transplantation and reliance on dual therapy with cyclosporin and azathioprine to provide adequate immunosuppression. This theoretically attractive option has, however, been associated with a greater than 50% incidence of acute rejection after cessation of steroid therapy. Any loss of renal function as a result of such a rejection episode may be more disadvantageous than continued low dose maintenance steroid therapy.

Table 15.2   Paediatric immunosuppression protocols.

Standard triple therapy	
Cyclosporin	
>20 kg	14 mg/kg/day (divided doses)
20 to 40 kg	12 mg/kg/day (divided doses)
Azathoprine	1.5 mg/kg/day, decreasing if WCC $< 4.0 \times 10^3$ and stopping if $< 2.0 \times 10^3$
Prednisone	1.0 mg/kg/day (max = 20 mg) till day 7, decreasing to 0.5 mg/kg/day by day 30, and 0.25 mg/kg/day by 3 months
Sequential (quadruple) therapy	
ATG	15 mg/kg/day
or OKT3	2.5 mg/kg/day if <30 kg, 5.0 mg/kg/day if >30 kg, for 10 days or until serum creatinine <200 μmol/l (whichever is sooner)
Cyclosporin	Withhold till creatinine <200 μmol/l and commence at 12 mg/kg/day, and if not <200 μmol/l by day 10 commence at 6 mg/kg/day and increase azathioprine dose to 2.5 mg/kg/day
Azathioprine, as for triple therapy	
Prednisone, as for triple therapy	
Target 12 hour trough whole blood cyclosporin levels (RIA)	
0–4 weeks	200–400 ng/ml
4–12 weeks	150–300 ng/ml
1 year	75–150 ng/ml

Sequential therapy

Prophylactic use of ALG or OKT3 maximises early immunosuppression and produces excellent graft survival. Acute graft rejection is usually delayed until after cessation of the antilymphocyte preparation, thus reducing management difficulties in the early postoperative period. This may be particularly important in the young child where primary graft function is of utmost importance. Nevertheless, ALG and OKT3 are expensive and associated with significant increases in viral infection and malignancy. For recipients of first cadaver donor kidneys, there is no advantage in terms of graft survival or graft function at one year after transplantation, which suggests that the use of sequential immunosuppressive therapy should be selective.

Increased immunosuppressive potency is probably required in young children less than six years of age, because of increased non-specific cellular immunological reactivity and propensity toward rejection. Additionally, it is this group of patients in which a fluid management is made difficult when initial or early graft function is poor. Highly sensitised children, and those receiving regrafts, may also benefit from sequential therapy. The remaining paediatric transplant recipients may be best served by use of a triple therapy protocol.

Early management after transplantation

For the first days after transplantation, the aim is to achieve primary graft function and early mobilisation of the child without complications of rejection, fluid overload, respiratory failure and technical mishaps. Achieving an uncomplicated early transplant period for the small child requires meticulous attention to fluid and electrolyte replacement.

Child weighing less than 20 kg.

The young child will return from the operating suite to an intensive nursing care area with a multilumen central venous catheter for fluid replacement and central venous pressure measurement, and an arterial line in the brachial artery for blood pressure monitoring. If an adult donor kidney has been placed in a very small child, ventilatory support may be required to overcome impaired respiratory movement produced by the sudden increase in intra-abdominal pressure. A resuscitation kit should be at hand with an appropriate sized endotracheal tube available, since the respiratory status of the uraemic child can be easily compromised by fluid overload and basal atelectasis that may follow inadequate pain control or prolonged ileus.

Intravenous fluids are administered to match the urine output together with an additional amount of 10 ml/kg/24 hours to replace insensible loss, in the ratio of two parts normal saline to one part 5% dextrose. Composition of intravenous solutions may require frequent adjustment depending upon results of serum and urinary electrolyte determinations performed at six hourly intervals during the first two days after transplantation. Additional potassium, sodium, calcium and bicarbonate are frequently required, and hyperglycaemia requiring insulin is common if large volumes of dextrose solutions are administered.

For the child passing urine at greater than 20 ml/kg/hour, it may be necessary to adjust the volume of replacement fluid more frequently than at hourly intervals. Adequacy of replacement therapy is best assessed by evaluating multiple clinical criteria rather than any one factor such as the CVP alone. Small children may become hypotensive due to fluid overload, despite reasonable levels of CVP, because of the relative preponderance of right ventricular failure. Attempts should be made to obtain a maximum circulating fluid volume without producing volume overload, particularly in the presence of large capacitance veins in an adult donor kidney. In such instances, an inadequate circulating volume may lead to thrombosis. Prolonged ileus may necessitate nasogastric drainage and parenteral nutrition, especially if the child was nutritionally deficient prior to transplantation.

The larger child

In the larger and older child, and depending on their underlying health status, it is more appropriate to manage fluid replacement according to CVP measurement and the previous hour's urine output, maintaining a CVP between 5 and 10 cm of water.

Infection

Bacterial urinary tract infections are common, occurring in more than half of paediatric renal transplant recipients, particularly in those with urinary tract abnormalities and previous urinary tract surgery. In such instances, low dose antibiotic prophylaxis is indicated. Primary cytomegalovirus (CMV) infection is common since exposure to CMV in children is uncommon prior to transplantation. It is therefore advantageous to avoid transplanting a kidney from a CMV positive donor into a CMV negative recipient, particularly if the recipient is to receive prophylactic ALG or OKT3. Chickenpox is a major problem in immunosuppresed children, as is shingles in those already infected with herpes zoster virus. Protection after known exposure can be provided by zoster immune globulin given . within 72 hours, and followed by regular observation for three weeks. If vesicles appear, azathioprine should be discontinued and parenteral acyclovir adminstered. Occasionally, disseminated infection producing encephalitis, carditis, pneumonitis and enteritis can result in death (Figure 15.4a & b).

Growth and development

Within the first two years of life, the child reaches half their adult height and thereafter grows at approximately 5 cm per year until puberty when the rate doubles for a further five years. The degree of growth retardation correlates

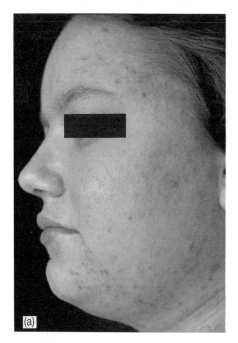 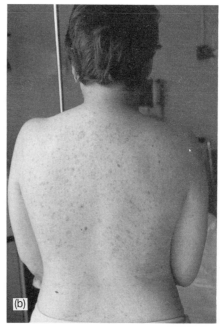

Figure 15.4 (a & b) This 19 year old girl, a recipient of a renal transplant five years earlier, developed chickenpox after contact with her younger sister. She subsequently developed disseminated infection and died two days after photography.

with the age of onset of renal sufficiency and the need for replacement renal therapy. Consequently, 70% of children who present for dialysis under the age of five years are under the third percentile for height, compared to 38% aged between 11 and 15 years. Children are now taller on reaching endstage renal failure because of better management before endstage, but rarely grow at a normal rate once on dialysis.

Bone age at the time of commencing treatment for renal insufficiency is a significant factor in determining the eventual adult stature. Bone age increases with increasing chronological age, with only minimal increase in height during dialysis treatment. Therefore, catch-up growth after transplantation is unlikely in children with a bone age of greater than 12 years, particularly if using steroid medication. In more than a third of instances, the adult height of long term paediatric renal transplant recipients receiving conventional immunosuppression with azathioprine and prednisone is at least two standard deviations below the mean. Causes of poor growth after renal transplantation include loss of growth potential during periods of dialysis treatment, poor renal transplant function and continued need for corticosteroid therapy, even in low doses.

Improved growth rates after transplantation may follow eventually as a result of better medical management of renal insufficiency. This includes avoiding a period of dialysis before transplantation and the use of cyclosporin with its steroid sparing effect. However, there is usually a gradual decline in renal function after the first year of transplantation, probably a result of chronic rejection and chronic cyclosporin nephrotoxicity. Furthermore, many children with renal transplants are dependent on corticosteroid therapy which depresses calcium absorption, decreases levels of 25-hydroxycalciferol and possibly increases levels of somatomedin inhibitors. As little as 2.5 mg of prednisone can prevent adequate growth hormone response, therefore suggesting that steroids have a direct inhibiting effect on growth hormone release.

An innovative approach to the renal transplant recipient who is growing poorly is to use exogenous recombinant growth hormone. Promising results have been demonstrated in renal transplant recipients with bone age less than 12 years and stable renal transplant function. Concern remains that exogenous growth hormone may precipitate acute rejection episodes, which must thus be watched for whenever growth hormone is used. Whereas physical growth is often disappointing after transplantation, the potential for catch-up of intellectual development is remarkable and may be the result of improved nutritional status, alleviation of uraemia and improved aluminium excretion.

Non-compliance

Graft loss or reduction in graft function as a result of rejection following non-compliance with immunosuppression is a major problem in paediatric renal transplantation, particularly in adolescent female patients. It is estimated that two thirds of paediatric recipients greater than 12 years of age exhibit non-compliant behaviour and more than half of the recipients of multiple renal transplants are non-compliant with their immunosuppressive therapy. One of the benefits of monitoring blood cyclosporin levels is the ability to detect non-compliance during periods of graft dysfunction. Non-compliant patients usually express concern about their outward appearance, cushingoid features, obesity, short stature, acne and hypertrichosis. Persisting non-compliance is more commonly seen in dysfunctional adolescents with chaotic families and minimal family support. Better strategies for achieving and monitoring immunosuppression compliance are urgently needed, particularly to overcome the frequently adverse effect of peer group assessment of paediatric renal transplant recipients.

Further reading

1. Sommerauer JF. Brain death determination in children and the anencephalic donor. *Clin Transplant* 1991; **5**, 137–145.
2. Keown PA, Lirenman D, Landsberg D *et al*. Use of cyclosporin in paediatric transplantation: Immunology, pharmacology and therapuetic implications. *Clin Transplant* 1991; **5**, 181–185.
3. Ettenger RB, Rosenthal JT, Marik J *et al*. Cadaver renal transplantation in children: Longterm impact of new immunosuppressive strategies. *Clin Transplant* 1991; **5**, 197–203.
4. Tejani A, Ingulli E. Growth of children post-transplantation and methods to optimise post-transplant growth. *Clin Transplant* 1991; **5**, 214–218.
5. Frawley JE, Farnsworth RH. Adult donor kidney transplantation in small children: A surgical technique. *Aust NZ J Surg* 1990; **60**, 911–912.

6. USRDS Annual Data Report 1991: ESRD in children. *Am J Kidney Dis* 1990; **18**(suppl.2), 79–88.
7. McEnery PT, Stablein DM, Arbus G, Tejani A. Renal transplantation in children. *N Engl J Med* 1992; **326**, 1727–1732.
8. Najarian JS, Frey DJ, Matas AJ *et al*. Renal transplantation in infants. *Ann Surg* 1990; **212**, 353–367.
9. Colon EA, Popkin MK, Matas AJ, Callies AL. Overview of non-compliance in renal transplantation. *Transplant Rev* 1991; **5**, 175–180.
10. Harman WE, Jabs K. Special issues in pediatric renal transplantation. *Semin Nephrol* 1992; **12**, 353–363.
11. Churchill BM, Steckler RE, McKenna PH *et al*. Renal transplantation and the abnormal urinary tract. *Transplant Rev.* 1993; **1**: 21–34.

16 Diabetic nephropathy and transplantation

Diabetes, a word of Greek origin meaning 'siphon' and indicative of the polyuria and polydipsia with which hyperglycaemic patients first present, is a complex metabolic disorder with variable long term sequelae. Until the discovery and clinical use of insulin in the mid 1920s, diabetics perished rapidly from the gross metabolic disturbances of dehydration produced by the osmotic effect of elevated circulating glucose levels, and ketoacidosis, the byproduct of cellular metabolism of alternative energy sources. A decade after the clinical introduction of insulin therapy, Kimmelstiel and Wilson first described the clinical syndrome of proteinuria, hypertension and progressive renal failure associated with nodular glomerular sclerosis (Figure 16.1). This syndrome of diabetic nephropathy, occurring in between 30 and 40% of chronic diabetics, appears at a variable time interval after diagnosis of diabetes and is the single most important cause of renal failure

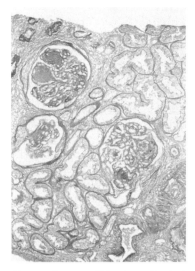

Figure 16.1 Renal cortex biopsy from a diabetic patient demonstrating characteristic nodular glomerulosclerosis, known as the Kimmelsteil-Wilson lesion (PAS, x32).

in Europe and the United States. Before introduction of replacement renal therapy in the form of dialysis and transplantation, renal failure was the commonest cause of death in insulin dependent diabetics. The numerical importance of diabetic nephropathy and the spectrum of available transplant management options warrant separate discussion in a manual of renal transplantation.

Diabetes mellitus

Diabetes is characterised by hyperglycaemia resulting from either a lack of production of insulin or the poor response of target cells to circulating insulin. The incidence varies between countries and racial groups, affecting approximately 3 to 5% of the population of most Western countries, and is low in Asia but higher in Scandinavia.

Type I diabetes

The more classical form of diabetes is associated with an absence of circulating insulin and affects about 10 to 15% of diabetics. This form is variously described as Type I diabetes, insulin dependent diabetes mellitus (IDDM) or juvenile onset diabetes, because in most instances, the condition is diagnosed before the age of 20 years. The lack of insulin is most likely the effect of an auto-immune process selectively destroying the insulin producing beta cells within the million or so islets which comprise the endocrine portion of the pancreas. The initiating event and development of the autoimmune response against the beta cell autoantigens are poorly understood, although circulating auto-antibodies in Type I diabetics include those directed against beta cell surface antigens and insulin. It is most commonly seen in individuals with DR3 and DR4 MHC class II antigens. Because of the large number of beta cells, the endocrine portion of the pancreas has a considerable functional reserve with elevated blood sugar levels not being observed until more than 80% of beta cells have been destroyed by the auto-immune process. Use of immunosuppressant therapy at the time of clinical diagnosis has been disappointing, principally because the onset of hyperglycaemia occurs at an irreversible stage of the disease process.

For most Type I diabetics, the disease is principally one of inconvenience. It can, however, be life threatening if insulin therapy is persistently inadequate, leading to hyperglycaemic ketoacidosis. Equally, a relative excess of insulin after prolonged vigorous exercise or missed meals can result in hypoglycaemic coma. In some insulin dependent diabetics, the combined deficiencies of glucagon and adrenaline secretory responses to hypoglycaemia lead to lack of awareness of hypoglycaemic symptoms of tachycardia and sweating, particularly in the early morning sleeping hours.

Type II diabetes

The majority of diabetics, often with a family history of the condition, develop hyperglycaemia in middle or old age (maturity onset) and are described as having Type II diabetes, which is characterised by hyperglycaemia in the presence of measurable circulating insulin. The pathogenesis includes a heterogeneous group of disorders such as defective beta cell insulin secretion, and resistance of target cells (e.g. liver, muscle) to insulin, particularly in obese patients. The resulting hyperglycaemia is usually controlled by dietary restriction and the use of oral hypoglycaemic agents which can both sensitise the beta cell to the stimulatory effect of glucose and increase sensitivity of target cell insulin receptors. Type II diabetics are not prone to develop ketoacidosis and are often referred to as having non-insulin dependent diabetes mellitus (NIDDM). Some, however, do require exogenous insulin, either as a result of chronic overstimulation of the beta cells by hyperglycaemia, or reduction of beta cell mass.

Long term complications of diabetes

The major concern of all diabetics of both types is development of chronic complications that may either lead to death from cardiovascular disease and renal failure or debilitating symptoms produced by damage to retinae and nerves. Although the pathogenesis of the two types of diabetes is different, the complications are similar, supporting the theory that the common problems of metabolic dysfunction and high blood sugar levels are the cause of the long term complications of diabetes. This may be an oversimplification, but nevertheless, the goal for treatment of diabetes is to normalise carbohydrate metabolism and maintain normal blood sugar levels. This aim is supported by mounting objective evidence, both experimental and clinical, that good glycaemic control both delays the onset and modifies the progress of the secondary complications. Glycaemic control may be assessed by measurement of blood glycosylated haemoglobin levels (HbA_{1c}) which should be less than 1.5 times the normal limits, a level that is rarely achieved with parenteral insulin. Although complications are more common in patients with poor glycaemic control, they still occur in those with good control. Unfortunately, predicting at an early stage which patients will ultimately develop these complications is not yet possible.

The development of secondary complications usually follows one of a number of recognisable clinical patterns. For example, diabetic nephropathy severe enough to cause renal failure usually occurs within 30 years of the diagnosis of Type I diabetes in patients less than 50 years of age, and is almost invariably associated with both severe proliferative retinopathy and neuropathy. Type I diabetics have a mortality rate about five times greater than the general population and develop renal failure ten times more commonly than Type II diabetics. Conversely, foot ulceration due to peripheral vascular disease and cataract formation is more frequent in Type II diabetics who have a mortality rate about twice that of the general population.

Natural history of diabetic nephropathy

The clinical phase of evidence of damage to the kidney is heralded by appearance of protein in the urine (> 200 mg/24 hours) which may initially be intermittent but is eventually present consistently. The peak incidence for diagnosis is between 15 and 30 years after clinical diagnosis of Type I diabetes. Once established, the rate of deterioration can be modified, but is rarely halted. Nephropathy in Type II diabetics is less well documented because the duration of glucose intolerance is usually unknown. Nevertheless, it is less common, probably progresses more slowly, but for several possible reasons is becoming numerically more important in renal replacement therapy programmes (Figure 16.2).

Preclinical phase

In the recently diagnosed Type 1 diabetic, there are changes in both function and appearance of the kidney. Apart from glycosuria and polyuria, the kidneys are increased in size as a result of hypertrophy and

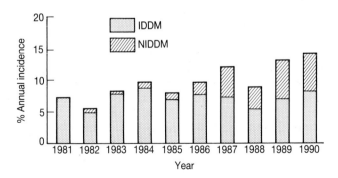

Figure 16.2 Changing pattern of presentation of diabetics accepted for replacement renal therapy in Australia from 1981 to 1990. Annual incidence of diabetes as a percentage of all causes of endstage renal failure.

hyperplasia, the GFR is increased and micro-albuminuria may be detected by sensitive assays. It is at this stage that progression towards established diabetic glomerulopathy may be avoided by near perfect glycaemic control; maintenance of blood pressure at levels no higher than normal for age; avoidance of tobacco; and reduction of animal protein in the diet, in addition to avoidance of refined carbohydrates, fat and alcohol. However, in a subset of diabetics and more commonly in those with poor glycaemic control, abnormal micro-albuminuria (>45 mg/24 hours) again develops and may be the earliest marker of the risk of progression to endstage renal disease, particularly in the presence of hypertension or a decreasing GFR.

Clinical phase

The clinical phase is usually preceded by a rise of blood pressure (due to sodium retention) within the generally accepted normal range. Hypertension may be the major cause of acceleration of the glomerulopathy. The time interval between the onset of proteinuria and establishment of chronic renal failure with the need for replacement renal therapy is about five years on average, but can vary from a few months to more than a decade. Diabetic retinopathy occurs more commonly than nephropathy and is present in more than 90% of diabetics at the time of presentation with proteinuria. The relationship between retinopathy and nephropathy is such that if proteinuria should occur in the absence of retinopathy in a patient with long standing diabetes, other causes of renal disease should be sought by renal biopsy. Peripheral neuropathy is common in patients with diabetic renal disease with severe symptoms resulting from the combined effect of uraemia and diabetes. Similarly, the gastro-intestinal symptoms of diabetic autonomic neuropathy are compounded by uraemic symptoms of anorexia, nausea and vomiting.

Progression of renal disease is often associated with marked deterioration in retinopathy as a result of hypertension and the effect of uraemia on haemostasis. The rate of progression of diabetic nephropathy can be reduced by good blood pressure control, perhaps particularly with angiotensin converting enzyme inhibitors, and perhaps by dietary protein restriction and cessation of smoking. The benefit of strict glycaemic control on established diabetic nephropathy and declining GFR is not proven.

Before widespread use of replacement renal therapy in the 1960s, renal failure was the most common cause of mortality in Type II diabetics. Even then, 80% of diabetics placed on haemodialysis died within the first year of treatment. Many dialysis centres were perhaps understandably reluctant to treat diabetics with endstage renal disease. With increasing sophistication of dialysis equipment and introduction of continuous ambulatory peritoneal dialysis in the late 1970s, the first year survival, in committed centres, increased during the 1980s to more than 80%. Nevertheless, diabetics on dialysis continue to do worse than non-diabetics because of continued progression of extrarenal complications.

Dialysis access

Vascular access in the diabetic patient may be difficult. Suitable arm veins are often sparse as a result of multiple venous cannulations in the years before the onset of diabetic nephropathy, while atheromatous plaque formation in arteries is frequent. Peritoneal dialysis is therefore an attractive option for diabetic patients requiring replacement renal therapy. The incidence of peritoneal infections is no different from that observed in non-diabetics and there is the potential of intraperitoneal administration of insulin. Greater amounts of insulin are required because of adsorption to plastic, poor absorption with loss in the drained dialysate and the glucose content of the peritoneal dialysate. However, delivery of insulin through the portal venous system, together with the intraperitoneal glucose reserve, provides greater stability of blood sugar levels than can be achieved by haemodialysis and subcutaneous insulin.

Histopathology

Overt morphological abnormalities other than enlargement of the glomerulus, are usually not apparent by light microscopy in the first years after the diagnosis of Type II diabetes. Electron microscopy may demonstrate thickening of the glomerular basement membrane within three or four years. Although the basement membrane may increase up to ten times in width, there is little correlation with the functional parameters of proteinuria, hypertension and decreased GFR. However, correlation does occur between these functional indices and expansion of the mesangium produced by accumulation of eosinophilic staining matrix material. As the mesangium expands, it compromises the lumen of glomerular capillaries. The nodular form of glomerulosclerosis is the most specific of the glomerular changes associated with diabetes, but is less common than diffuse glomerulosclerosis. Other renal changes include diffuse intimal fibrosis of large renal vessels, hyalinisation of efferent arterioles, atrophy of tubules and interstitial fibrosis.

Other diabetic urinary tract pathology

In addition to nephropathy, diabetics are commonly afflicted by other renal tract abnormalities that frequently serve further to compromise renal function and complicate renal transplantation.

Cystopathy

Urinary tract infections are probably not more common in diabetics when compared to the non-diabetic population, except in a subgroup of diabetics who develop diabetic cystopathy caused by impairment of bladder sensation as a result of diabetic autonomic neuropathy. The condition is characterised by infrequent voiding, a feeling of incomplete emptying, straining and a poor urinary stream. Urodynamic studies demonstrate a large bladder capacity, impaired bladder sensation, impaired urinary flow rate, lowered detrusor muscle pressure and increased residual volume after voiding. At cystoscopy, there is no evidence of anatomical outflow obstruction.

Most patients with diabetic nephropathy have no symptoms of diabetic cystopathy but the majority, if tested by urodynamic studies, will have evidence of the condition. The earliest signs are the presence of a residual volume and urinary tract infection, both of which are potentially hazardous to a renal transplant. Hence, early recognition of urinary tract infection and appropriate treatment are important. The use of anticholinergic drugs to increase detrusor muscle function is of questionable benefit because of the frequent absence of cholinesterase activity in the diabetic bladder. Bladder outflow resistance can be improved by alpha blockade using prazosin which has few side effects. Transurethral sphincterotomy is not recommended because of resultant incontinence. Intermittent clean urinary catheterisation and maintenance antibiotic therapy is a practical option and should spare most patients from the need for diversion of the upper urinary tract into a bowel conduit.

Papillary necrosis

This condition probably follows infarction of a renal papilla, although urinary tract infection is often present at the time of necrosis. Presentation may range from asymptomatic haematuria to oliguric renal failure in combination with an infected and obstructed upper urinary tract. The diagnosis can be made by intravenous pyelography or CT scanning with radiographic vascular contrast. Radiographic contrast should be used with care as it may induce acute and sometimes irreversible renal failure, particularly in diabetic patients. Adequate hydration and use of non-ionic contrast followed by an infusion of mannitol may reduce the incidence of this complication. Treatment is usually supportive, except in the presence of infection, when drainage or even nephrectomy may be required.

Comprehensive management of diabetics with nephropathy

Diabetic patients, particularly those with poor glycaemic control, are prone to ignore their medical problems. The need for regular and frequent blood sugar monitoring and insulin injection is at best an inconvenience, but replacement renal therapy is a major psychological and social disturbance to the diabetic patient and their family. Retinopathy should be well controlled by prophylactic and therapeutic laser retinal photocoagulation and symptoms of autonomic and peripheral neuropathy are usually compatible with an acceptable lifestyle. When these problems are compounded by the cachexia and lethargy of uraemia, patients with diabetic nephropathy frequently lose independence and self-esteem. They seem to be overcome by their mounting medical problems. Their family life is stressed by the frequent inability to maintain employment and for the male, by the additional loss of sexual activity because of the adverse effects of autonomic neuropathy on penile erection. Emotional support for the patient with developing or established diabetic nephropathy is therefore often difficult to achieve.

Establishment of an integrated programme is vitally important to provide comprehensive and optimal management of the diabetic patient, both before and after developing diabetic nephropathy, and should include coordinated active involvement of diabetologists, paediatricians, nephrologists and ophthalmologists, together with dialysis and transplant services. Such a programme will prolong the preclinical and clinical phases of diabetic nephropathy, allow for gradual acceptance of the limitations imposed by renal replacement therapy and provide realistic expectations of the values of both renal and pancreatic transplantation. Patients' support groups are of additional benefit, particularly in association with regular education sessions by medical and nursing staff.

Assessment for transplantation

Diabetics account for between 10 and 40% of patients presenting for replacement renal therapy, depending on factors such as their country of origin and age limitations. Many diabetic dialysis dependent patients will not be suitable for renal transplantation because of the high incidence of co-existing cardiovascular pathology. Nevertheless, and even though no randomised prospective trials comparing renal transplantation and dialysis for replacement renal therapy in patients with diabetic nephropathy have been published, transplantation is arguably the best therapeutic option. Restrictive recipient selection for transplantation would provide better results in terms of patient and graft survival for a given transplant centre, but precludes many diabetics from the best form of treatment of their renal failure.

About 40% of deaths in most series of diabetic dialysis and transplant patients are a result of myocardial infarction or congestive cardiac failure. The diabetic heart is not only vulnerable to coronary artery disease, but also to fluid overload and the added effect of uraemia on cardiac muscle function. The combination of smoking, diabetes and renal failure is even more troublesome, so serious consideration should be given to not including current smokers on the transplant waiting list. All diabetic patients who are to be considered for transplantation require evaluation to exclude significant coronary artery disease. Those with symptomatic cardiac disease warrant investigation by coronary angiography. Thallium radionucleotide scanning of the myocardium before and after stress provided by exercise or persantin may be used to screen for coronary artery disease. Significant proximal coronary artery disease, irrespective of the presence of symptoms, should be treated by coronary artery bypass grafting or angioplasty before the diabetic patient is placed on the transplant waiting list.

In Australia, where the incidence of Type I diabetes is low and no upper age limit exists for acceptance onto a dialysis programme, 60% of diabetic patients presenting for replacement renal therapy have Type I diabetes (Figure 16.3). The remaining Type II diabetics are older and less likely to be placed on a transplant waiting list. Diabetics are seldom given preference on cadaver transplant waiting lists, and whilst waiting, often develop co-existing pathology that subsequently prevents transplantation. Many transplant centres overcome this problem by placing emphasis on living related transplantation before the need for dialysis.

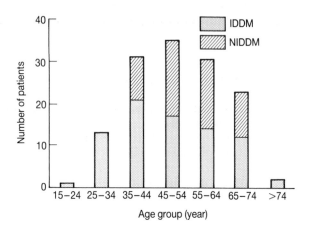

Figure 16.3 Distribution of age of diabetic patients presenting for replacement renal therapy in Australia in 1990. IDDM = insulin dependent diabetes mellitus. NIDDM = non insulin dependent diabetes mellitus.

Renal transplantation

Initial experiences with renal transplantation for the treatment of diabetic nephropathy in the 1960s were disappointing and many centres did not offer diabetics the option of renal transplantation. Those centres that did reported separate results for non-diabetic and diabetic recipients, since graft survival was detrimentally affected by transplantation into a diabetic recipient. Although the results of renal transplantation in the diabetic recipient have improved dramatically over the last two decades, when compared to transplantation in the non-diabetic recipient, differential patient and graft survival of between 10 and 20% at three years continues to exist, with graft loss as a result of patient death from cardiovascular disease being frequent.

The University of Minnesota transplant group, 40% of whose patients have diabetes as a cause of renal failure, have reported 32% graft survival in diabetics ten years after transplantation, comparable to those of non-diabetic recipients in other centres. Although they have demonstrated that long term graft survival was feasible, their diabetic patients continued to have evidence of progression of neuropathy, retinopathy and large vessel disease. Graft loss as a result of diabetic nephropathy, however, did not occur in the first ten years after transplantation and was minimal thereafter.

The surgical procedure of renal transplantation and the current initial immunosuppressive regimens are the same as those used in non-diabetic recipients. Corticosteroids usually make glycaemic control more difficult and the advent of cyclosporin, with its steroid sparing effect, has therefore been beneficial. It has thus been possible carefully and slowly to reduce or even withdraw the maintenance steroid treatment. When intravenous steroid bolus treatment of acute rejection is used, an insulin infusion provides the best form of glucose homeostasis. Alternatively, steroids can be avoided by using biological preparations to treat episodes of acute rejection.

Rationale for pancreas transplantation

Transplantation of insulin producing cells would be an appropriate therapy for insulin dependent diabetes if it could be achieved without substantial risk of morbidity or mortality to the recipient. Providing normal or near normal self-regulated glucose homeostasis by transplantation would resolve the disturbed lifestyle and impaired quality of life caused by insulin dependent diabetes, avoid the risk of recurrent hypoglycaemic events in patients with defective glucagon mediated counter-regulatory responses and, perhaps most importantly, prevent appearance or progression of secondary complications of diabetes.

There is now convincing evidence from animal studies and increasing data from long term follow-up of diabetics with advanced complications to suggest that pancreas transplantation is capable of preventing progression of secondary complications of diabetes. Pancreas transplantation of all diabetics, however, is both unnecessary and impractical, as the majority of diabetics do not develop secondary complications or have only minor problems, and there are insufficient allogeneic cadaver pancreas donors to transplant even a small proportion of the diabetic population. Nevertheless, until such time as large numbers of diabetics either in the preclinical or early clinical phase of diabetic nephropathy receive a pancreas transplant alone, the true value of pancreas transplantation may not be seen. In the meantime, the well documented improved quality of life may be sufficient indication to continue with pancreas transplant programmes, particularly in diabetics already in need of a renal transplant. Current indications for vascularised pancreas transplantation in patients with Type I diabetes are listed in Table 16.1.

Combined pancreas and kidney transplantation

Simultaneous transplantation of the pancreas and kidney from the same donor is a technically viable treatment option for diabetic nephropathy. With careful patient selection, one year pancreas graft survival in major centres is greater than 80%, a figure that is comparable with other forms of solid organ transplantation. For this group of patients, in whom renal transplantation is otherwise necessary, there is no additional risk to the patient as a result of immunosuppression, and kidney graft survival is not compromised by additional pancreas transplantation. In such instances, the optimum timing of the transplant procedure would be before the diabetic patient becomes significantly affected by the nutritional and metabolic effects of uraemia, perhaps when the GFR has declined to 20 ml/min.

The remaining question is whether the undoubted benefits of simplification of diabetes management are worth the risk of technical complications of the pancreas transplant procedure. Opinion among transplant centres and diabetologists is divided, but the consensus view is that pancreas transplantation is justified in centres undertaking comprehensive clinical research programmes, even though these patients often have advanced and irreversible complications of diabetes at the time of transplantation. Pancreas transplantation after previous successful renal transplantation is also a valid therapeutic option as the diabetic patient is already immunosuppressed.

Pancreas transplantation alone

The indications for pancreas transplantation alone are continuing to evolve with clearly defined guidelines yet to be established. In major centres the graft survival for pancreas transplants in non-uraemic diabetics is between 60 and 70% at one year, with rejection being the cause of increased pancreas graft loss when compared to the simultaneous transplant procedure (Figure 16.4). Continued evaluation of the pancreas transplantation alone procedure is justified, at least in major centres, where the indications for transplantation include aggressive neuropathy, frequent hypoglycaemia and resistance to subcutaneous insulin. Available evidence suggests that retinopathy is, however, best managed by laser photocoagulation.

The majority of pre-uraemic patients considered for pancreas transplantation alone have biopsy evidence of diabetic nephropathy and hence, maintenance of remaining native renal function is of major concern. Renal function is further adversely affected by the nephrotoxic effect of cyclosporin, often halving the GFR within a month of transplantation. A conservative acceptable GFR before transplantation is 70 ml/min. Early data, which should be treated with some caution, suggests that renal function is maintained after the initial decline, perhaps providing evidence of the beneficial effect of normal glycaemic control on the progression of diabetic nephropathy.

Table 16.1 Current indications for vascularised pancreas transplantation in patients with type I diabetes mellitus

Indication	Transplantation
1. Uncomplicated diabetes	No
2. Retinopathy	No
3. Early nephropathy	No
4. 'Brittle' diabetes	? Pancreas alone
5. Progressive autonomic neuropathy	Pancreas alone
6. Subcutaneous insulin resistance	Pancreas alone
7. Hypoglcyaemia unawareness	Pancreas alone
8. Endstage renal disease	Pancreas and kidney
9. Advanced cardiovascular disease	No

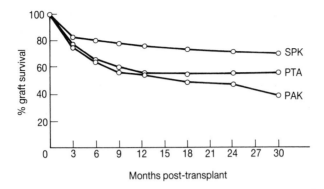

Figure 16.4 Actuarial pancreas graft functional survival rates by recipient category (SPK = simultaneous pancreas and kidney, *n* = 833; PTA = pancreas alone, *n* = 71; PAK = pancreas after kidney, *n* = 112) reported to the United Network for Organ Sharing Registry of USA patients from October 1987 to October 1990 (reproduced with permission from *Clin Transplant* 1991; **5**, 330–341).

Vascularised pancreas transplantation

The first clinical pancreas transplant was performed in Minneapolis in December, 1966. The pancreas graft functioned for two months, a similar fate to many other pancreas transplants performed over the next 15 years that failed because of either uncontrolled rejection or problems associated with surgical technique. However, the 1980s saw an exponential growth of pancreas transplantation and more than 800 procedures are reported each year to the International Pancreas Transplant Registry. This growth is a reflection of improved graft survival provided by the use of cyclosporin, better patient selection and surgical advances.

Surgical technique

Much of the history of pancreas transplantation relates to the variety of techniques required to overcome the central problem of unwanted exocrine drainage from the pancreas. Duct ligation and free drainage into the peritoneal cavity proved to be failures. Duct occlusion with a synthetic polymer and subsequent transplantation of the body and tail of the pancreas was the first major technical advance, despite the common complications of pancreatic fistulae following incomplete occlusion and graft thrombosis because of diminished blood flow requirement. Transplantation of the whole pancreas, rather than a segment, however, maximises the number of available islets and is now possible with either enteric or bladder exocrine drainage.

More than 80% of the pancreas transplants performed in recent years have involved the relatively simple bladder drainage technique described by the University of Wisconsin transplant group in 1983. The whole pancreas is retrieved in continuity with the second part of the duodenum from donors aged between ten and 45 years. Preservation with University of Wisconsin solution allows the pancreas to be stored in ice for periods of up to 24 hours. The whole pancreas, together with the adjacent second part of the duodenum stapled and oversewn at both ends, is placed in an intraperitoneal position on the right side through a midline incision. The vascular anastomoses are to the external iliac vessels, with exocrine bladder drainage provided by anastomosis of the antimesenteric border of the transplanted second part of the duodenum to the dome of the urinary bladder (Figure 16.5).

Arterial blood supply to the pancreas is provided via an aortic patch containing splenic and gastroduodenal artery branches of the coeliac trunk and the inferior pancreaticoduodenal branch of the superior mesenteric artery. Venous drainage via the transplanted portal vein into the external iliac vein provides systemic insulin delivery. If the liver is also retrieved from the cadaver donor, as is the usual case, the aortic patch, coeliac trunk and most of the portal vein are unavailable to the pancreas transplant surgeon, necessitating use of donor iliac vessels as extension grafts (Figure 16.6). Although the vascular reconstruction is tedious, the subsequent arterial anastomosis placed obliquely across the recipient common iliac artery bifurcation is technically easier than use of a long aortic patch.

Some pancreas transplant centres prefer enteric drainage of the pancreatic exocrine juices to a loop of small bowel. Although providing physiological exocrine drainage, it is associated with more local complications than the bladder drainage technique, perhaps because the anastomosis cannot

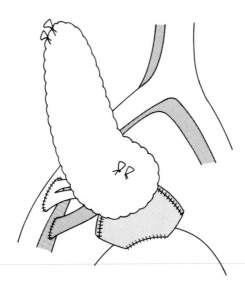

Figure 16.5 Vascularised whole pancreas transplant with urinary bladder exocrine drainage via a segment of donor duodenum.

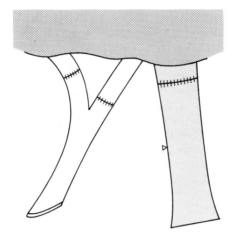

Figure 16.6 Reconstruction of the pancreaticoduodenal graft vasculature after concomitant donor liver retrieval, using donor iliac vessels to replace the donor portal and provide a common arterial supply to the splenic and superior mesenteric arteries.

be protected by the equivalent of a urinary catheter.

Monitoring rejection

Assessment of pancreas transplant endocrine function by measurement of blood glucose levels is a poor marker of pancreas rejection because of the large functional reserve provided by the islets. A major advantage of the bladder exocrine drainage technique is the ability to measure amylase excretion in the urine. Within a 24 hour period, amylase excretion into the urine will vary considerably according to oral intake and circulating gut hormone levels. However, over a 24 hour period, falls of greater than 25% may be significant, but not specific for rejection. The earliest and perhaps most specific marker for pancreas rejection is the implied diagnosis provided by the impaired function of the concomitant renal transplant and probably explains the improved pancreas graft survival when simultaneously transplanted with a kidney from the same donor. In the absence of a functioning renal transplant, particularly in the first two weeks after transplantation when the urinary amylase levels are low, percutaneous needle core biopsy (20 gauge needle) provides a reliable and safe diagnosis of rejection of the pancreas.

Complications

Venous thrombosis is responsible for loss of 5 to 10% of pancreas grafts, principally in the first week after transplantation. The pancreas transplant is prone to thrombose because of the relatively low blood flow through a large capacitance vein, potential technical problems associated with orientation of the portal vein anastomosis, and propensity for thrombosis of small veins within the pancreas graft if preservation has been poor. Most pancreas transplant centres therefore use some form of prophylaxis against thrombosis, usually subcutaneous heparin.   Urinary tract infections are more common than in diabetic patients receiving kidney transplants alone, but of more concern is the activation of pancreatic digestive pro-enzymes by urea splitting bacteria. These enzymes are probably responsible for the late onset of intermittent haematuria, cystitis and urethritis that occurs between two and six months after transplantation in about 40% of patients. These problems often resolve with treatment of the urinary tract infection and bladder protection by temporary urinary catheterisation. In some instances, it is necessary to divert the exocrine drainage to a loop of small bowel, a procedure that does not appear to have the same frequency of complications as creation of enteric drainage as a primary procedure. Nevertheless, these problems indicate that the technique of bladder exocrine drainage for pancreas transplantation is less than perfect.

Glucose homeostasis

Vascularised pancreas transplantation undoubtedly provides simplification of diabetic management such that most recipients can rightly describe themselves as former diabetics. Near perfect control of blood glucose levels can be achieved (Figure 16.7) even though about half the patients have fasting hyperinsulinaemia and abnormal insulin responses to an oral glucose challenge (Figure 16.7). Despite fasting hyperinsulinaemia in 50% of patients 12 months after transplantation, the fasting triglyceride levels in the Westmead series of patients fell significantly to normal levels when compared to those before transplantation, while fasting serum cholesterol levels did not change. The possible causes of the fasting hyperinsulinaemia include peripheral resistance to insulin provided by steroid medication, non-physiological systemic delivery of insulin and denervation of the pancreas.

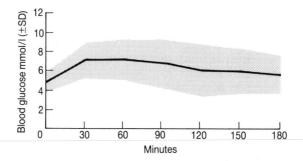

Figure 16.7   Mean (± SD) blood glucose response to 75 gm oral glucose challenge, one year after transplantation, in 16 consecutive patients receiving technically successful bladder drained pancreas transplants at Westmead Hospital.

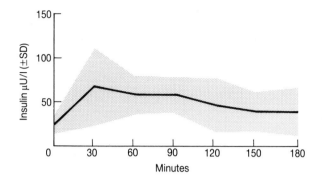

Figure 16.8 Mean (± SD) biphasic serum insulin response to 75 gm oral glucose challenge, one year after transplantation, in 16 consecutive patients receiving technically successful bladder drained pancreas transplants at Westmead Hospital.

Isolated islets

Transplantation only of the endocrine portion of the pancreas is appealing, especially as implantation of separated islets into a suitable portal venous drainage site would not require complex surgery. After many years of evaluation in experimental models, limited reports of successful clinical isolated islet transplantation first appeared in 1990. Because of difficulties associated with separation and isolation of human islets and rejection, success has been short term and often dependent on using up to five pancreas donors. In the short term, it therefore seems unlikely that isolated human islet cell transplantation will become a practical option for diabetics. The pig may, in the future, become an important source of islet tissue. Xenotransplantation across species barriers may become an important source of islet tissue, such that diabetics may again come to rely on porcine insulin delivered directly by functioning pig islet transplants.

Further reading

1. Dubernard JM, Sutherland DER. *International Handbook of Pancreas Transplantation.* Dordrecht: Kluwer Academic, 1989.
2. Freidman EA. Comprehensive care essential to success in diabetic nephropathy. *Transplant Proc* 1984; **16**, 3–6.
3. Lane P, Steffes M, Mauer SM. The role of the paediatric nephrologist in the care of children with diabetes mellitus. *Pediatr Nephrol* 1991; **5**, 359–363.
4. Mogensen CE. Prediction of clinical diabetic nephropathy in IDDM patients. *Diabetes* 1990; **39**, 761–767.
5. Najarian JS, Kaufman DB, Fryde DS *et al.* Long term survival following kidney transplantation in 100 Type 1 diabetic patients. *Transplantation* 1989; **47**, 106–113.
6. Sutherland DER. Indications for pancreas transplantation: A commentary. *Clin Transplant* 1990; **4**, 242–246.
7. Sutherland DER, Gillingham K, Moudry-Munns KC. Results of pancreas transplantation in the United States for 1987–90 from the United States Network for Organ Sharing (UNOS) registry with comparison to 1984–87 results. *Clin Transplant* 1991; **5**, 330–341.
8. Sollinger HW, Knetchle SJ, Reed A *et al.* Experience with 100 consecutive simultaneous kidney–pancreas transplants with bladder drainage. *Ann Surg* 1991; **214**, 703–711.

9.   Tyden G, Tibell A, Groth CG. Pancreatico-duodenal transplantation with enteric exocrine drainage: Technical aspects. *Clin Transplant* 1991; **5**, 36–39.
10. Nankivell B, Allen RDM, Bell B *et al*. Factors affecting measurement of urinary amylase after bladder-drained pancreas transplantation. *Clin Transplant* 1991; **5**, 392–397.
11. Allen RDM, Wilson TG, Grierson JM *et al*. Percutaneous biopsy of bladder-drained pancreas transplants. *Transplantation* 1991; **51**, 1213–1216.
12. Rosen CB, Frohnert PP, Velosa JA *et al*. Morbidity of pancreas transplantation during cadaveric renal transplantation. *Transplantation* 1991; **51**, 123–127.
13. Allen RDM. Transplantation of the endocrine pancreas. *Med J Aust* 1992.

Appendix 1
Terminology of transplant immunology

This appendix is designed to provide a brief description of the main terms used in immunology and transplantation, but it is not intended as a comprehensive guide to transplantation immunology.

Adhesion molecules
Molecules present on the surface of cells which function as links to bind cells involved in the immune response. There are a number of families of adhesion molecules, where the members of each family have a similar structure. Examples include integrins, very late antigens (VLA), selectins, addressins and intracellular adhesion molecules (ICAMs).

Table A1: Adhesion molecules involved in transplant responses

Molecule	Cell	Binding to	On	Reason
LFA-1	Leukocytes	ICAM	Endothelium	Leucocyte adhesion
LFA-3	Endothelium	CD2	T lymphocyte	Non-specific cell activation
VLA-4	Leucocytes	VCAM-1	Endothelium	Leucocyte
VLA-5	Leucocytes	Fibronectin	Extracellular matrix	Matrix binding

Allo-
Allo-antigen, allogenic, allograft. These terms use 'allo' to describe a relationship between a recipient and an antigen, gene or graft respectively from a donor, in which the two are of the same species but not genetically identical (see *Syngeneic* and *Xenogeneic*).

Antibody
A molecule with a highly variable region which defines specificity for a particular antigen, binds to its target and activates phagocytic cells and complement to mediate destruction of the target (see *Immunoglobulin*).

Anti-idiotypic antibody
This is a particular type of antibody with a variable region which is specific for binding to the variable region of another antibody molecule.

Antigen presenting cell (APC)
Specialised cells which process antigens and present them in the binding groove of HLA molecules, to naive or sensitised T lymphocytes. There are two pathways for presentation of antigen. Exogenous proteins, present in the external environment of the cell, are endocytosed and bound to class II HLA molecules. Endogenous proteins are processed by intracellular proteasomes designed to deliver 'bite sized' peptides to newly formed class I HLA molecules, which are then expressed on the cell surface complete with the peptide. Dendritic cells and macrophages are the predominant antigen presenting cells.

Antilymphocyte globulin (ALG) (Antithymocyte globulin, ATG)
Antibodies produced by an animal (horse, goat or rabbit) that has been injected with human lymphocytes or thymocytes respectively. Serum from the immunised animal is then purified for use in prophylaxis or treatment of rejection (see Chapter 7).

Autologous
Present on a person's own cells or in their own serum.

Auto-antibody
An antibody that reacts with the person's own cells or molecules. It may be cytotoxic or non-cytotoxic.

B lymphocyte
B lymphocytes (B cells) are a subgroup of lymphocytes which differentiate in bone marrow. Each B lymphocyte has a specific immunoglobulin gene arrangement which produces a specific immunoglobulin molecule. A process of 'allelic exclusion' leads to production of only one heavy and one light chain. Maturation of the B lymphocyte leads to expression of HLA class II molecules, CD19, 20, 21 and the Fc receptor, as well as the immunoglobulin specific to that B lymphocyte. Activation, by binding of the immunoglobulin to antigen, leads to clonal proliferation, followed by secretion of immunoglobulin by the resultant plasma cells. 'CD5 positive' B lymphocytes are amongst the earliest B lymphocytes generated in neonatal life. These lymphocytes, which bear the surface molecule CD5, produce immunoglobulins that bind carbohydrate molecules present on bacteria, but also crossreact with a number of antigens present on autologous cells. These are generally regarded as primitive lymphocytes because of their lack of fine specificity.

Beta$_2$ microglobulin
A constant molecule found associated with class I HLA molecule alpha chains, completing the structure of class I molecules expressed on the cell surface.

CD nomenclature
CD (Cluster Determinant) molecules are present on the surface of cells and

have been used to distinguish specific subsets of lymphocytes with particular functional characteristics. Monoclonal antibody production techniques led to recognition of many distinct molecules present on the surface of lymphocytes and other cells. The CD nomenclature scheme was devised in order to control the proliferating and diverse nomenclature that was building up in the early 1980s. A number of international workshops have been used to define and then refine the nomenclature. Examples of the molecules and their function are shown in Table A2.

Table A2: Examples of Cluster Determinant molecules

CD	Present on	Function
CD2	T lymphocyte	Receptor binding sheep erythrocyte and LFA-3
CD3	T lymphocyte	Essential element for T lymphocyte receptor function
CD4	T lymphocyte subset	Binds HLA class II molecules, also receptor for HIV
CD8	T lymphocyte subset	Binds HLA class I molecules
CD11a	Leucocytes	Alpha chain for LFA-1
CD11b	Monocytes, granulocytes	Alpha chain for complement receptor
CD18	Leucocytes	Beta chain for CD11a, b
CD28	Leucocytes	Second signal binds B7
CD45	Leucocytes	Leucocyte common antigen
CD58	Leucocytes, epithelium	LFA-3
CD68	Macrophages	Marker for macrophages

Complement
A cascade of serum proteins, manufactured in the liver, having important effector functions in the immune system. Like the clotting cascade, the complement system amplifies the response to a single triggering event. The similarity extends to there being two mechanisms of stimulating complement activation and the existence of a number of control mechanisms which prevent overactivation. The 'classical' complement pathway is triggered by binding of immune complexes to C1q. The 'alternative' pathway is triggered by bacterial surface polysaccharides which bind one of the complement proteins and protect it from inactivation by normal control proteins. Both mechanisms lead to conversion of complement factor C3a to C3b. Production of C3b leads to a series of effector molecules which cause chemotaxis; increase in local vascular permeability; activation and granule release from mast cells, granulocytes and macrophages; and formation of membrane attack complexes which lead to lysis of target cells.

Complement dependent cytotoxicity (CDC)
In transplantation, the term CDC is used to describe the standard crossmatch test in which lysis of lymphocytes occurs when recipient antibody binds to donor lymphocytes which are subsequently incubated with rabbit complement.

Cytokines

Cytokines are a series of molecules secreted by cells of the immune system which act locally, either in the cell which produced them (autocrine), or on other cells in the vicinity (paracrine). The cytokine system is complex, with the balance between competing cytokines important in regulating many immune functions. Different T lymphocyte subsets may affect the immune response in different ways in response to a single stimulus. Table A3 shows a number of cytokines and a limited selection of known actions.

Table A3: Cytokines and their major actions

	Main Secretory Cell	**Main target and functions**
Interleukins		
IL1	Macrophage	Activated lymphocyte proliferation
IL2	T lymphocytes	Lymphocyte activation and proliferation
IL3	T lymphocytes	Differentiation of haemopoietic stem cells
IL4	T lymphocytes	B lymphocyte activation
IL5	T lymphocytes, macrophages	B lymphocyte proliferation and differentiation
IL6	Macrophages, fibroblasts, T lymphocytes	B lymphocyte differentiation
IL7	Bone marrow	Lymphocyte proliferation
IL8	Fibroblasts, macrophages	Chemotaxis
IL9	T lymphocytes	T lymphocyte proliferation
IL10	T lymphocytes, macrophages	Down regulate class II and cytokine production
Tumour necrosis factors		
TNF alpha	Macrophages, T lymphocytes	Tumour necrosis
TNF beta	T lymphocytes (CD4 +ve)	Acute phase response, phagocyte activation
Interferons		
IFN alpha	Leucocytes	Increase HLA class I expression
IFN beta	Fibroblast	Antiviral
IFN gamma	T lymphocytes	Increase HLA expression
Colony stimulating factors		
GM CSF	T lymphocytes, macrophages	Growth and activation of granulocytes and macrophages
M CSF	Fibroblasts, endothelium	Growth of macrophages
G CSF	Fibroblasts, endothelium	Growth of granulocytes

Cytotoxic T lymphocyte (CTL)

These T lymphocytes usually bear the surface marker CD8, which binds to a non-polymorphic section of HLA class I molecules. They are directly toxic to the cells which they bind, through release of tumour necrosis factor and perforins. Thus a cell infected with virus, and expressing virus peptide in the antigen presenting groove of HLA class I, will be bound by the cytotoxic T cell through its T cell receptor and CD8 molecule and subsequently killed.

Dendritic cell

The dendritic cell is a specialised antigen presenting cell which has the function of accumulating foreign antigen, processing it and presenting it to T lymphocytes in association with HLA class II molecules. The classic example is the Langerhans cell in the skin, which is thought to collect antigen and migrate to regional lymph nodes to activate T lymphocytes. In the context of a renal transplant, the donor's dendritic antigen presenting cells are capable of presenting directly to recipient cells. The indirect route, by contrast, requires donor antigen to be processed by recipient dendritic cells before presentation (see also *Antigen Presenting Cells*).

Epitope
Each molecule has its own defined 3-dimensional structure and part of that 3-D structure may be recognised by a particular antibody. The exact configuration that is bound directly to an antibody is the 'epitope'. A large molecule may have many epitopes, each detected by a specific antibody, leading to polyclonal recognition of the molecule.

Fc region and Fc receptor
The constant portion of an immunoglobulin is termed the Fc or constant fragment to distinguish that end of the molecule from the highly variable portion at the other end. The cell surface of monocytes, macrophages, polymorphs, neutrophils and eosinophils has a specific binding receptor (Fc receptor) to which the Fc portion of immunoglobulin binds. It can thus be seen that the Fc portion acts as a link between an immunoglobulin and a functional cell. For example, immunoglobulin binds a parasite invader with the variable portion and is then bound via the Fc portion to the Fc receptor on an eosinophil, which can thus release cytotoxins locally and kill the parasite.

Flow cytometry
Flow cytometry is the principle behind most cell counting equipment used in automated measurement of blood counts. The essential concept is that cells are entrained singly through a flow chamber. Having thus arranged for cells to be channelled one at a time through a predetermined position, it is possible to use laser light beams and photomultiplier tubes to measure size and count numbers. The principal application of flow cytometry in transplant immunology is to use fluorescent cell surface markers which can be activated by laser light at a particular frequency and emit light at a different frequency. It is thus possible to use monoclonal antibodies to 'tag' fluorescent markers onto particular cells, such as those expressing the CD8 molecule, and thus to count their number.

Graft-versus-host (GVH)
GVH is the response of lymphocytes present in the donor graft against the recipient or host tissues. The most frequent and severe form of graft-versus-host disease occurs in bone marrow transplantation because large numbers of viable lymphocytes are transplanted with the marrow. Graft-versus-host reactions also occur after multiple blood transfusions and occasionally after renal transplantation. The best recognised GVH in renal transplant recipients is due to plasma cells in the graft producing antibodies which react with recipient erythrocytes causing haemolytic anaemia. Recent discovery of long lived donor cells in recipient skin, lymph node and liver has raised the possibility that some form of subclinical GVH reaction may occur in most patients.

H-2
The H-2 system is the mouse equivalent of the human leucocyte antigen (HLA) system; both are names for the major histocompatibility complex genes in the respective species. There is a tendency for experimental and clinical scientists to speak different languages, the former using H-2, the latter HLA nomenclature. H-2 genes in the mouse produce either class I molecules (H2K, H2D, H2L) or class II molecules (IA, IE). Instead of using numbers to designate the particular molecule such as HLA-B27 the mouse system uses a lower case character, e.g. H-2K^d (see HLA, MHC).

Helper T lymphocyte
The 'helper' subset of T lymphocytes are characterised by possession of the CD4 molecule on the cell surface. The CD4 molecule binds to a non-polymorphic region of the class II molecule. Helper T lymphocytes recognise antigen in the peptide binding groove of class II molecules and provide 'help' to B lymphocytes and cytotoxic T lymphocytes. Some uncertainty still remains about whether or not helper cells are an essential prerequisite for graft rejection in the clinical setting.

HLA
Human leucocyte antigens (HLA) are the molecules which define most of the immune response to a foreign tissue or organ transplant. They are identified by tissue typing and derived from genes in the major histocompatibility complex (see Chapter 5 and *MHC*).

Ia
Mouse class II major histocompatibility complex gene (see *H-2*).

Idiotype
Immunoglobulins vary in amino acid sequence in their V region. These variations in the immunoglobulin can themselves act as antigens which other immunoglobulins can recognise and bind to. The parts of the V regions to which anti-idiotypic antibodies bind are termed idiotopes.

Immunoglobulin
A major component of serum protein, the immunoglobulins/antibodies are a group or family of molecules produced by B lymphocytes. The molecules may be bound to the cell membrane or secreted into blood, other extracellular fluids, or body secretion. Five major antibody classes (IgG, IgM, IgA, IgD and IgE) and a number of subclasses (e.g. IgG2) are distinguished by the structure of their heavy chain component. Two types of light chain (kappa and lambda) combine with all of the types of heavy chain to complete the antibody molecule. Antibodies directed at the HLA molecules are responsible for the 'humoral' component of graft rejection.

Interferons
A group of molecules first detected by their ability to prevent viral infection of cells exposed to them before virus. One of the families of cytokine with a potential role in graft rejection (see *Cytokines*).

Interleukins
A family of molecules characterised by their role of messengers between leucocytes – hence their name. IL1 and IL2 are central to the process of graft rejection (see Chapter 7 and *Cytokines*).

Isograft
Transplant between two genetically identical individuals, for example identical monozygotic twins.

LFA
Lymphocyte function antigen (see *Adhesion molecules*).

Lymphocytes
A series of mononuclear cells responsible for specific defence of the individual against invading micro-organisms. Also responsible for many 'allergic' and 'auto-immune' diseases as well as transplant rejection. Lymphocytes are broadly divided into T and B lymphocytes, each with specific maturation pathways and functions (see *T-lymphocytes, B-lymphocytes, CD*).

MHC
Major histocompatibility complex (MHC) is the term used to describe the genetic region (on the short arm of chromosome 6 in man) which encodes the molecules that provide the major stimulus to graft rejection. There are at least two reasons why the products of the MHC have proved the major barrier to transplantation. Firstly they are probably the group of proteins that vary most from individual to individual and are thus recognisable as foreign. Secondly the molecules are responsible for the central event in the individual's immune defence against virus – recognition of foreign antigen. The MHC region of man is described more fully in Chapter 5. It encodes class I molecules, class II molecules, and a series of molecules described as class III molecules. The class I molecules comprise a polymorphic alpha chain which combines with $beta_2$ microglobulin. Class II molecules have both alpha and beta chains encoded in the MHC. The ultimate structure of class I and class II molecules is similar, with a peptide binding groove on the exposed upper surface, into which peptides are bound for presentation to receptors on the surface of T lymphocytes.

MLR/MLC
Mixed lymphocyte reaction/culture is a test in which lymphocytes from two individuals are mixed *in vitro*. The subsequent response is proliferation, if the lymphocytes originate from individuals with different class II molecules. Cellular proliferation can be measured by the uptake of 3H-thymidine into cells. The MLR has been used as a test for determining genetic identity at class II for bone marrow transplantation. It has largely been superseded by DNA typing techniques for bone marrow transplants and discarded in renal transplantation.

NK cells
Natural killer cells are large granular lymphocytes with the capacity to lyse target cells without prior exposure. The normal target *in vivo* is a cell infected with virus, but the effects of NK cells are also seen *in vitro* and may be relevant to transplant rejection.

Oligonucleotide

Oligonucleotides are short sequence of nucleotides created *in vitro* and designed to mimic one strand of DNA or be complementary to a strand of RNA. Oligonucleotides are usually in the order of 20 nucleotides in length and can be designed as a starting point to create replication of a particular region of DNA (see *PCR*) or as a probe to bind to, and thus detect, a specific region of DNA (see *SSP/SSOP*).

PCR

The polymerase chain reaction (PCR) is the most powerful tool of modern genetics, with applications extending to the detection of small quantities of any form of DNA or RNA in almost any situation. At the extreme it is possible, using the PCR technique, to determine the tissue type of an individual from the cells clinging to the root of an extracted hair.

   The principle of PCR is to determine an area of gene that one wants to multiply from small numbers of copy up to detectable quantities of cDNA, and then to use a cyclic process of gene amplification. In this way a small amount of virus DNA, such as that of cytomegalovirus, can be multiplied many millions of times such that it becomes easily detectable by routine biochemical tests. The main problem is that the PCR is exquisitely sensitive and subject to producing false positive results from a very small amount of contamination. Application of PCR to tissue typing has provided a rapid and accurate method of typing the DNA of an individual. It has thus taken tissue typing from assessment of the cell surface molecule to assessment of the genetic code for those molecules (see *SSP/SSOP*).

Proteasome

This is an intracellular protein, the gene for which is encoded in the class II region of the MHC, responsible for a variety of different proteolytic processes. It is thought that proteasomes are responsible for cutting proteins into suitable sized peptides for association with newly produced MHC class I molecules.

RFLP

Restriction fragment length polymorphism is a technique for determining differences in segments of DNA, but it is largely being replaced by the PCR technique (see *PCR*). The essential element of the RFLP technique is a 'restriction enzyme' which is capable of cutting DNA at a particular point by recognising a particular nucleotide sequence. There are many different restriction enzymes recognising and cutting at different sequences. It has thus been possible to design tests which use restriction enzymes to cut DNA differently depending upon the tissue type of an individual and therefore to use the test as a method of DNA based tissue typing.

Second set response

A second set response is defined as a rejection response that occurs on the second exposure of a recipient to tissue from a particular donor. A skin graft between two mice will be rejected several times faster on the second compared with the first occasion. Migration of polymorphonuclear leucocytes and lymphocytes into the graft with subsequent graft destruction is much more rapid in the second set response. This represents the response of a primed or sensitised, rather than an immunologically naive, individual.

SSP/SSOP

'Sequence specific primers' and 'Sequence specific olignucleotide probes' are two alternative ways of using the PCR technique (see *PCR*) to tissue type. The principle is essentially the same for each, namely a region of the class II gene of interest (e.g. DRB1) is amplified and then detected using an oligonucleotide probe with a radioactive or enzymatic marker attached. In the SSP technique the initial 'primers', which are the oligonucleotides used to amplify the segment of gene, are specific for a particular tissue type and the product is detected by a non-specific oligonucleotide. In the SSOP technique, the reverse is true with the region of interest amplified in all individuals but then detected using oligonucleotides that are specific to particular tissue types. SSP is quick, but cannot detect very refined differences in tissue type, while SSOP is slow and expensive to apply but produces very specific definition of the class II tissue types.

Syngeneic

A transplant between syngeneic individuals occurs when they have the same genetic type. Identical monozygotic twins provide the only syngeneic human transplant. Inbred strains of laboratory animals provide an experimental model for investigating differences between syngeneic and allogeneic (see *Allo*) transplants.

T lymphocyte

The T lymphocyte (T cell) is one of the major subsets of lymphocytes, distinguished by its cell surface molecules and by its function. T lymphocytes develop in the thymus where they are 'educated', firstly by positive selection for T lymphocytes which only recognise self HLA molecules and secondly by negative selection, whereby those that recognise self antigens in the peptide binding groove of self HLA molecules are eliminated. The T lymphocyte thus recognises self HLA molecules via its T cell receptor (see *TCR*) but may have one of many alternative functions (see *Cytotoxic T Cell*, *Helper T Lymphocyte*).

TCR

The T lymphocyte/cell receptor is a heterodimer on the cell surface made up from two chains (alpha and beta chains usually, gamma and delta chains more rarely). The receptor acts on the T lymphocytes' recognition structure for binding to the individual's HLA molecules – either class I or class II. Following binding the TCR transmits an intracellular signal by calcium influx controlled by the CD3 molecule.

Tumour necrosis factor (TNF)

TNF is one of the families of cytokines (see *Cytokines*) release of which is probably responsible for the 'first dose' symptoms that result from administration of OKT3.

Tolerance

A state claimed by immunologists to describe the situation in which a transplant 'should be' rejected but is not. These situations demand explanation and hence the use of the word tolerance. The mechanisms whereby tolerance is achieved vary from animal model to animal model. Tolerance in clinical practice remains the 'Holy Grail' which all are striving to find.

Xeno

Xenogeneic/xenograft implies transplantation between two different species, e.g. pig to dog or mouse to rat. Two different situations are encountered: concordant xenografts between closely related species, where there are no naturally formed pre-existing antibodies that react to the donor (e.g. monkey and man); and discordant, where the recipient has naturally occurring antibodies to the donor species (e.g. pig and man). Some believe that xenotransplants will dramatically change the face of clinical transplantation within ten years.

Appendix 2
Cyclosporin drug interactions

Drug	Major or substantiated					Minor or infrequent					Comments
Generic class name	Increased Cy levels	Increase nephrotoxicity	Decrease Cy levels	Changes interacting drugs' dose/level	Other	Increased Cy levels	Increase nephrotoxicity	Decrease Cy levels	Changes interacting drugs' dose/level	Other	
Acetazolamide						+	+				
Acyclovir						+	+				
Alcohol											
Amiloride combinations						+					
Aminoglycocides		+									
Amphotericin B		+			+						
Androgenic steroid	+										Cy dose should increase up to 50%
Anticonvulsants			+								
Atacurium							?				Potentiates neuromuscular blockade
Calcium channel antagonist						?					See individual drugs; most, not nifedipine
Captopril							?				
Carbemazepine			+								
Cefotaxime							?				
Ceftazidine							?				
Cefuroxime							?				
Cephradine							?				
Chlorpropamide					+		?				May lead to antabuse like reactions to alcohol
Cimetidine						?	?				
Ciprofloxacin						?	?				
Colchicine	+										
Corticosteroids (high dose iv)		+									
Cortrimoxazole (high dose iv)		+									
Danazol	+										In some patients
Diclofenac		+									

Drug	Major or substantiated					Minor or infrequent					Comments
	Increased Cy levels	Increase nephrotoxicity	Decrease Cy levels	Changes interacting drugs' dose/level	Other	Increased Cy levels	Increase nephrotoxicity	Decrease Cy levels	Changes interacting drugs' dose/level	Other	
Dicoumarin						+	?		?		
Digoxin											
Diltiazem	+										Cy dose should drop 30–50%
Disopyramide											
Diuretics		+					?				Probably via salt loss
Dopamine										+	Decreases nephrotoxicity
Doxorubicin											
Doxycline											
Erythromycin	+										Cy dose should drop 50%
Fluconazole	+						+				Individual variation
Frusemide											
Gentamicin		+									
H₂ antagonists						+	?				
HMG Co A reductase inhibitors					+						Rhabdomyolysis
Imipenem											
Indomethacin		+									Acute renal failure
Isoniazid			+								Manageable by Cy dose increase
Itraconazole	+										
Ketoconazole	+										Cy dose should drop 30–50%
Levonorgestral						+					
Lovastatin					+						Rhabdomyolysis
Macrolide antibiotics	+										See erythromycin
Mannitol											
Melphalan		+					?				

Drug	Major or substantiated					Minor or infrequent					Comments
	Increased Cy levels	Increase nephrotoxicity	Decrease Cy levels	Changes interacting drugs' dose/level	Other	Increased Cy levels	Increase nephrotoxicity	Decrease Cy levels	Changes interacting drugs' dose/level	Other	
Metamizole						+			+	+	
Methotrexate						+					Increase methotrexate levels
Methylprednisolone						+					In some patients
Methyltestosterone											
Metoclopramide	+										
Metolazone											
Metoprolol											
Metronidazole	+					+					
Naficillin			+								
Nicardipine	+						+				
Nifedipine						+				?	Possible increase in gingival hypertrophy
Norethisterone											
Norfloxacin	+					?					
NSAIDs		+									
Octreotide			+			?		?			
Omeprazole											
Oral contraceptives	+										
Pancuronium				+							Potentiates neuromuscular blockade
Pentazocine											
Phenobarbitone			+								Manageable by Cy dose increase
Phenytoin			+								Manageable by Cy dose increase
Posinomycin	+										
Prednisolone	+		+		+			+			Variable, may decrease prednisolone clearance
Primidone			+								
Probucol			+								

Drug	Major or substantiated					Minor or infrequent					Comments
	Increased Cy levels	Increase nephrotoxicity	Decrease Cy levels	Changes interacting drugs' dose/level	Other	Increased Cy levels	Increase nephrotoxicity	Decrease Cy levels	Changes interacting drugs' dose/level	Other	
Propafenone	+							?			Disagreement in literature
Ranitidine											
Rifampicin			+		?						Almost unmanageable. May need 500% Cy dose increase
Simvastatin											Possible rhabdomyolysis
Sulindac						+		+			
Sulphinpyrazone	+					+		+			
Tamoxifen	+										
Thiazide diuretics		+					+				
Tobramycin		+									
Ticarcillin	+										
Trimethoprim		+									
Vancomycin											
Vecuronium			+		+ +						Potentiates neuromuscular blockade
Warfarin	+										Requires both drugs to be monitored carefully

+ = Definite interaction
? = Conflicting or doubtful evidence only

Source: Ann Templeman, Sandoz, Australia

Index

autoreactive 61–62
endothelial-monocyte 63
Fc 289
historical positive crossmatch 59,65
idiotypes 286, 290
IgG, *see* IgG, IgM,
lymphocytotoxic 58–64
non-HLA 63
panel reactivity 60–61, 65
Anticoagulation, *see* heparin
Anti-fungal agents 230–232
Antigen presenting cells, *see also* dendritic
cells, passenger leukocyte 91
Antiglomerular basement membrane
disease 9–10
Antigens, *see also* ABO system
alloantigens and the immune
response 90–92
antibodies 58–60
crossmatching 58–64
HLA system 43–44, 53–57, 66–67, 90–92
immune response to an allograft
rejection 65, 90–92
minor histocompatibility, *see* minor
histocompatability antigens
presentation 51–53, 91–92, 288
presenting cells 90–92, 286, 288–289
Anti-idiotypic antibodies 286
Antilymphocyte serum/globulin (ALS/ALG),
see also antithymocyte globulin 99–100,
105–108
administration 100–101
children 263
chronic rejection 174
clinical results 100
cytomegalovirus 101, 221–222
history 99–100
malignancy 100
mechanisms of action 100
protocols 100, 262
rejection treatment 105–108, 151–152
side effects 100, 101
Antithymocyte globulin (ATG) 100–102
Anti-viral agents 223–224
ANZDATA registry 6, 57, 66
Aortic patch 76, 77, 152, 260
Appendicitis 260
Arterial complications, *see* renal artery
Arteriography, *see* angiography
Arteriosclerosis, *see* atherosclerosis
Aseptic bone necrosis, *see* avascular bone
necrosis
Aseptic meningitis, with OKT3 101, 102
Aspergillosis 231
Atherosclerosis, *see* vascular disease
Autoantibodies, *see also* antibodies,
crossmatching, HLA system 61–62,
286
Autologous, definition 286
Avascular bone necrosis 99, 107, 209, 211,
245–246

Avascular ureteric necrosis 155–157
Azathioprine
allopurinol interaction 96
complications 97–98
cyclosporine and 98, 105
dialysis 98
donor specific transfusions, *see* blood
transfusion
drug interactions 96
haematological effects 96–97
hair loss 97, 98
hepatic dysfunction 15
historical perspective 2, 96
leucopaenia 96, 249
maintenance therapy 97, 105–106
malignancy 241–244
mode of action 96
pharmacology 96
pregnancy 253
recommended regimen 96, 97, 105–106
dose 96, 97, 105–106
side effects 96–98
toxicity 96–98, 249
triple therapy 98

B cells, *see also* antibodies 53–54, 286
crossmatching 60, 61
Bacterial infection, *see* infection
Balloon dilatation
artery 215
ureter 212–213
Barrier nursing 218–219
Basal cell carcinoma 242–243
Beta-2-microglobulin 286
Biopsy, *see* fine needle aspiration biopsy,
transplant biopsy
Bladder abnormalities
children 12, 256
clots 159
diabetes mellitus 273–274
leak 156, 204, 208, 213–214
pre-transplant assessment 12
washout 141
Blood groups, *see* ABO system, Lewis
system
Blood transfusion 16–20
cyclosporin treated patients 16
donor specific 18–19
during surgery 74
graft survival rates 16–17
historical perspective 16
living related donor transfusion 18–19
preparation 16–17
pre transplantation 70
post transplantation 115
results 17, 19
sensitization risk 17
storage time 17
Bone disease, *see also* avascular
necrosis 98–99, 209, 211, 245–247
children 256